# Spon's Estimati
# Minor Works, A                                              rs
# to Fire, Flood, G                                   ﹍﹍age

# Spon's Estimating Costs Guide to Minor Works, Alterations and Repairs to Fire, Flood, Gale and Theft Damage

Fourth edition

**Bryan Spain**

Taylor & Francis
Taylor & Francis Group
LONDON AND NEW YORK

First published 1999 by Spon Press
Second edition 2004
Third edition published 2006 by Taylor & Francis
This edition published 2009 by Taylor & Francis
2 Park Square, Milton Park, Abingdon, Oxon OX14 4RN

Simultaneously published in the USA and Canada
by Taylor & Francis
270 Madison Avenue, New York, NY 10016, USA

*Taylor & Francis is an imprint of the Taylor & Francis Group,
an informa business*

*Publisher's Note*
This book has been prepared from camera-ready copy supplied by
the author.

Printed and bound in Great Britain by
CPI Antony Rowe, Chippenham, Wiltshire

*British Library Cataloguing in Publication Data*
A catalogue record for this book is available from the British Library

ISBN 10: 0–415–46906–6 Paperback
ISBN 13: 978–0–415–46906–7 Paperback

ISBN 10: 0–203–89484–7 eBook
ISBN 13: 978–0–203–89484–2 eBook

# Contents

Preface                                                    xi

Introduction                                               xiii

Standard Method of Measurement/trades link                 xxi

## Part One: Unit rates

Demolition, excavation and filling

Demolition                                                 3
Excavation                                                 5
Earthwork support                                          7
Disposal by hand                                           7
Disposal by machine                                        8
Filling by hand                                            9
Filling by machine                                         9
Surface treatments                                        10

Concrete work

Ready-mixed concrete                                      11
Site-mixed concrete                                       12
Formwork                                                  13
Reinforcement                                             17
Joints                                                    18
Concrete finishes                                         18
Precast concrete                                          18

Brickwork and blockwork

Brickwork                                                 20
Blockwork                                                 30
Damp proof courses                                        34
Sundries                                                  36

Masonry

Walling                                                   37
Sundries                                                  39

Carpentry and joinery

Sawn softwood, untreated                41
Sawn softwood, impregnated              42
Gutters and fascias                     44
Wrought softwood supports               45
Metal fixings                           46
Flooring                                48
Linings                                 49
Windows                                 50
Doors                                   52
Door frames and linings                 53
Stairs                                  55
Kitchen fittings                        56
Insulation                              57
Ironmongery                             57

Metalwork

Balustrades                             60
Lintels                                 60
Sundries                                61

Roofing

Tiling                                  62
Fibre-cement slating                    65
Natural slating                         66
Reconstructed stone slating             68
Lead sheet coverings                    69
Copper sheet coverings                  71
Built-up roofing                        72

Asphalt work

Damp proofing and tanking               74
Flooring                                75
Roofing                                 76

Floor, wall and ceiling finishings

Screeds                                 78
In situ wall coverings                  82
Lathing                                 84
Plasterboard                            85
Floor tiling                            86
Wall tiling                             88

Plumbing and heating

Sanitary fittings 90
Holes and chases 96
Rainwater goods 100
Waste systems 107
Soil pipes 110
Overflow systems 112
Traps 112
Copper pipework, capillary joints 113
Stop valves 117
Copper pipework, compression joints 117
Cold water storage tanks 118
Hot water copper cylinders 119
Insulation 120
Boilers 120
Oil storage tanks 122
Radiators 123

Glazing

Clear float glass 126
White patterned glass 128
Georgian wired cast glass 130
Georgian wired polished glass 131
Clear laminated safety glass 132

Painting and wallpapering

On internal surfaces before fixing 134
On internal walls and ceilings 134
On internal woodwork 136
On internal metalwork 141
On external woodwork 145
On external metalwork 150
Wallpapering 155

External works

Drainage 157
Manholes 164
Fencing 166
Kerbs and edgings 168
Sub-bases 169
Concrete beds 170
Pavings 171

Alterations and repairs
    Shoring 173
    Forming openings 174
    Filling openings 176
    Underpinning 177
    Temporary screens 182

Spot items
    Brickwork, blockwork and masonry 183
    Roofing 186
    Carpentry and joinery 189
    Finishings 192
    Plumbing 193
    Glazing 195
    Painting 195
    Wallpapering 196

## Part Two: Damage repairs

Emergency measures 199
Fire damage 201
Flood damage 212
Gale damage 214
Theft damage 218

## Part Three: Approximate estimating

Excavation and filling 227
Concrete work 228
Brickwork and blockwork 229
Masonry 230
Carpentry and joinery 231
Finishings 234
Plumbing and heating 234
Painting 236
Wallpapering 237
Drainage 238

## Part Four: Plant and tool hire

Concrete and cutting equipment 243
Access and site equipment 244
Lifting and moving 246
Compaction 246
Breaking and demolition 246
Power tools 247
Welding and generators 248
Pumping equipment 248

# Part Five: General construction data

General construction data                                        251

# Part Six: Business matters

Starting a business                                             271
Running a business                                              287
Taxation                                                        293

Index                                                           303

# Preface

This the fourth edition of Spon's Estimating Costs Guide to Minor Works, Alterations and Repairs to Fire, Flood, Gale and Theft Damage and is intended to provide accurate cost data for contractors carrying out work in contracts valued up to £50,000.

The need for accurate estimates that are prepared quickly is vital for all contractors operating in the domestic construction market. Most contractors have the skills necessary to carry out the work together with the capacity for dealing with the setbacks that are part of the normal construction process. What they do not usually have is enough time to complete the many tasks that must be carried out in order to trade profitably. This book aims to help contractors to prepare their estimates by providing thousands of rates for small building work. If used sensibly, the book can save valuable time for contractors in the preparation of their bids and can also play a part in making sure that value for money is received from sub-contractors.

I have received a great deal of support in the research necessary for this type of book and I am grateful to those individuals and firms who have provided the cost data and other information. In particular, I am indebted to Mark Loughrey of Loughrey & Co Ltd, Chartered Accountants of Hoylake (tel: 0151-632 3298 or www.mjloughrey@accountant.com), who specialise in advising small construction businesses. Their research for the information in the business section is based on tax legislation in force in March 2008.

Although every care has been taken in the preparation of the book, neither the publishers nor I can accept any responsibility for the use of the information provided by any firm or individual. Finally, I would welcome any constructive criticism of the book's contents and suggestions that could be incorporated into future editions.

Bryan Spain
www.spainandpartners@btconnect.com
July 2008

# Introduction

This is the fourth edition of Spon's Estimating Costs Guide to Minor
Works, Alterations and Repairs to Fire, Flood, Gale and Theft Damage
and it follows the layout, style and contents of other books in this series.
The contents of the book cover unit rates, project costs, repairs, tool and
equipment hire, general advice on business matters and other information
useful to those involved in the commissioning and construction of
landscaping and external works. The unit rates section presents analytical
rates for work up to about £50,000 in value and the business section
covers advice on starting and running a business together with information
on taxation and VAT matters.

## Materials

In the domestic construction market, contractors are not usually able to
purchase materials in large quantities and cannot benefit from the
discounts available to larger contractors. An average of 10% to 15%
discount has been allowed on normal trade prices.

## Labour

The hourly labour rates for craftsmen and general operatives are based
upon the current wage awards. These are set at:

| | |
|---|---|
| Craftsman | £16.00 |
| General operative | £14.00 |
| Advance Plumber | £17.50 |

These rates include provision for NIC Employers' contribution, CITB levy,
insurances, public and annual holidays, severance pay and tool
allowances where appropriate.

## VAT

The costs displayed in this book exclude VAT.

## Headings

The following column headings have been used.

| Unit | Labour | Labour cost £ | Materials £ | O & P £ | Total £ |
|---|---|---|---|---|---|
| m2 | 0.20 | 3.20 | 2.20 | 0.78 | 5.98 |

## Unit

This column shows the unit of measurement for the item description:

| | |
|---|---|
| nr | number |
| m | linear metre |
| m2 | square metre |
| m3 | cubic metre. |

## Labour

In the example shown, 0.20 represents the estimated time estimated to carry out one square metre of the described item, i.e. 0.20 hours.

## Labour cost

The entry of £3.20 is calculated by multiplying the entry in the Labour column by the labour rate of £16.00.

## Materials

This column displays the cost of the materials required to carry out one square metre of the described item, i.e. £2.20.

## O & P (Overheads and profit)

This has been set at 15% and is deemed to cover head office and site overheads including:

- heating
- lighting
- rent
- rates
- telephones
- secretarial services
- insurances
- finance charges
- transport
- small tools
- ladders
- scaffolding etc.

## Total

This is the total of the Hours, Materials and Overheads and Profit columns.

## Contracting

Tradesmen and small contractors can act as main contractors (working for a client direct) or as a sub-contractor working for another contractor. Although a contract exists between a sub-contractor and a main contractor, there is no contractual link between a sub-contractor and an Employer.

In general terms this means that the sub-contractor cannot make any claims against the Employer direct and vice-versa. It also means that the sub-contractor should not accept any instructions from the Employer or his representative because this could be taken as establishing a privity of contract between the two parties,

A sub-contractor must be aware of his role in the programme because if he causes a contractor to overrun the completion date for the main contract he may become liable for the full amount of liquidated damages on the main contract plus the cost of damages that the contractor and other sub-contractors may have suffered.

A well-organised sub-contractor will keep a full set of daily site records, staffing levels, plant on site, weather charts and such like. It also cannot be over emphasised that any verbal instructions that the sub-contractor receives, should be confirmed immediately to the contractor in writing with the name of the person who issued them.

This procedure is extremely important because it may eventually save the sub-contractor considerable expense if someone tries to lay the blame for delays to the contract at his door. It is also important that instructions should only be taken from the contractor and he should be informed if another party attempts to do so.

## Contractor's discount

Most sub-contracts allow for a discount to the contractor of 2.5% from the sub-contractor's account. This means that the sub-contractor must add this discount to his prices by adding 1/39th to his net rates.

## Payment and retention

Payment is normally made on a monthly basis. The sub-contractor should submit his account to the contractor who then incorporates it into his own payment application and passes it on to the Architect or Employer's representative for certification of payment. When the sub-contractor receives his payment it will be reduced by 5% retention.

This money is held by the Employer and will be released in two parts. The first part, or moiety, is paid at the completion of work and, in the sub-contractor's case, this may be either when he has finished his work or when the contractor has completed the contract as a whole (known as practical completion) depending upon the contract conditions. The second part is released at the end of the defects liability period.

## Defects liability period

This is the period of time (normally 12 months) during which the sub-contractor is contractually bound to return to the job to rectify any mistakes or bad pieces of workmanship. This could either be twelve months from when he completes his work or twelve months from when the main contract is completed depending upon the wording of the sub-contract.

## Period for completion

Usually, a sub-contractor will be given a period of time in which he must complete the work and he must ensure that he has the capability to do the work within that period. Failure to meet the agreed completion date could have serious consequences.

Under certain circumstances, however, particularly with nominated sub-contracts, the sub-contractor may be requested to state the period of time he requires to do the work. If this is the case, then careful thought must be given to the time inserted. Too short a time may put him at financial risk but too long a time may prejudice the opportunity of winning the contract.

## Damages for non-completion

A clause is usually inserted within each sub-contract stating that the sub-contractor is liable for the financial losses that contractor suffers due to the sub-contractor's non-completion of work on time. This will include the amount of liquidated and ascertained damages contained within the main contract, together with the contractor's own direct losses and the direct losses of his other sub-contractors. As can be seen, the potential cost to the sub-contractor can be large so he must take care to expedite the work with due diligence to avoid incurring these costs.

## Variations

All sub-contracts contain a clause allowing the sub-contract work to be varied without invalidating it. The sub-contractor will normally be paid any additional cost he incurs in carrying out variations.

## Insurances

The sub-contractor is responsible for insuring against injury to persons or property and against loss of plant and materials. These insurances could be taken out for each individual job, although it is more common to take out blanket policies based on the turnover the firm has achieved in the previous year.

**Extensions of time**

The sub-contractor will normally be entitled to a longer period of time to complete the work if he is delayed or interrupted by reasons beyond his control (known as an extension of time). Most sub-contracts list the reasons and in some cases the sub-contractor may also be entitled to additional monies as well as an extension of time.

**Domestic sub-contracts**

In domestic sub-contracts the contractor would obtain competitive quotations from various sub-contractors of his own choice and these may be based on a bill of quantities, specification and drawings, or schedules of work. Accompanying the enquiry should also be a form of sub-contract that the sub-contractor will be required to complete.

There are several points that may affect costs and which the sub-contractor should bear in mind. These are

1.  Whether the rates and prices are to include for any contractor's discount (normally expressed as plus 1/39th allow 2.5%).

2.  Whether the contractor is to supply any labour or plant to assist the sub-contractor in either carrying out any of the work or in off-loading materials.

3.  What facilities (if any) the contractor will provide for the sub-contractor such as mess rooms, welfare facilities, office accommodation and storage facilities.

4.  Whether the contractor is to dispose of the sub-contractor's rubbish.

**Contracting**

Often a sub-contractor will find himself working under a private contract, written or implied. This usually takes the form of working for a domestic householder or a small factory owner and the following procedures usually apply in this type of work.

## Estimate

The initial approach would usually come from a purchaser, e.g. 'How much will it cost to have my garden landscaped?' At this stage, he may only want an approximate cost in order to see if he can afford to have the work carried out as opposed to a quotation which is a firm offer to do the work. Therefore, a brief description of the work to be carried out together with an approximate price will suffice.

However, it should be made clear that the price is an estimate and does not constitute an offer that may be accepted by the purchaser. The estimate may be based on a telephone conversation only, e.g. 'It will cost about £4,000 to £ 5,000 to landscape your garden', or it could be based on a brief visit to the house. In either case, little time should be spent on an estimate and it is generally wise to express it as a price range.

## Quotation

A quotation is generally seen as an offer to do the work for the price quoted, and could constitute a simple contract if accepted. It follows that some time and effort should be spent in compiling a quotation to save arguments at a later stage. One should always remember that the contractor is the expert and must use his expertise in order to guide the purchaser and should discuss the work with him in full. He should tell the purchaser exactly what he is getting for the price and also what he is not.

This may mean going in to some detail such as what will happen to the surplus excavated materials, how access will be gained, how long the job will take and similar items.

The contractor should also find out from the purchaser exactly what restrictions (if any) will be placed upon him. For instance, will the purchaser keep the drive clear of cars to allow a skip to be used and will the contractor only be allowed entry to the premises on certain days and/or at certain times? These factors, should be ascertained in advance, and the costs of complying with them should be made known to the purchaser who may decide to take steps to change the restrictions.

Once the contractor has considered all the relevant factors then the formal written quotation can be produced. It should state precisely what the purchaser is getting for his money, including when and how long the job will take and contain all the salient points of discussions that have taken place.

After a quotation has been submitted then all that needs to be done is for the purchaser to accept it. Although a verbal acceptance would constitute a binding agreement, it is always more satisfactory if the acceptance is made in writing.

## Payments

There is much debate on how and when payments should be made in domestic situations. Ideally from a contractor's point of view to be paid in advance would be the most advantageous, but the chances of the purchaser wishing to do this are remote.

On the other hand, it may cause undue financial hardship to a recently self-employed contractor to have to buy all the materials himself and not get paid until all the work is completed. Whatever payment policy is adopted it must be agreed with the purchaser in advance and form part of the written quotation.

Possible alternatives are

1. Being paid when the work is complete. This is probably the best method from a public relations aspect and contractors who can complete a job in a few days should have no difficulty in adopting this policy.

2. Being paid before the work is done. This is only really feasible where the contractor concerned is of unquestionable reputation or is well known to the purchaser.

3. Being paid for materials as they are bought and delivered with the balance paid when the work is complete. This could be a practical solution for smaller contractors, but the purchaser will probably want proof of the material costs, so careful handling of invoices is necessary.

4. Some form of stage payments that usually take the form of agreed percentages of the quotation price or agreed parts of the quotation price paid after stages of the work have been carried out.

## Pricing and variations

It is important that some method of recording, pricing and being paid for variations is agreed at the outset and this is particularly relevant when dealing with private clients. Unforeseen additions, more than any other item, are the main cause of disputes and are often avoidable.

The risk of this type of dispute can be reduced by ensuring that the original quotation is as detailed as possible. The detailed specification of the materials could be contained within the descriptions or done separately. A quotation broken down in this way is detailed enough to enable the purchaser to ascertain that he is not being overcharged for any variations that may occur and yet is not so detailed that the purchaser is going to question the price of every detail.

Also, if the purchaser should wish to change anything himself then there are no arguments on what was included in the original quotation.

If variations occur, it must be established who should pay for them. There are three main types of variations.

1. Those instructed by the purchaser.

2. Those that should have been included in the original quotation.

3. Those that are necessary due to events that could not have been foreseen.

The liabilities for 1 and 2 are relatively straightforward. If the purchaser says he wants a different paving flag to his original choice, then he must bear the additional cost. Conversely, if the contractor forgot to include the cost of the sub-base in his quotation then it is only fair that he bears the cost.

Item 3 is more difficult. If it is the purchaser who is receiving the benefit of the variations and if they were not foreseeable, then it would be logical to assume that it is the purchaser who should bear the cost. An example would be where the excavation to a patio revealed old foundations underneath, the contractor would expect to be paid the extra cost for removing them.

Other instances may not be as clear cut as this example and it may become necessary to arrive at a cost-sharing arrangement if genuine doubt exists. Variations should preferably be agreed in advance before the work is carried out. They should be recorded and signed by both parties and, wherever possible, priced in detail and agreed.

# Standard Method of Measurement/ Trades Links

The contents of this book are presented under trade headings and the
following list provides a link to the Standard Method of Measurement.

**Demolition, excavation and filling**

> C10 Demolishing structures
> D20 Excavation and filling

**Concrete work**

> E10 In situ concrete
> E20 Formwork for in situ concrete
> E30 Reinforcement for in situ concrete
> E40 Designed joints for in situ concrete
> E41 Worked finishes to in situ concrete
> E50 Precast concrete units

**Brickwork and blockwork**

> F10 Brick/block walling
> F30 Accessories for brick/block walling

**Masonry**

> F20 Natural stone rubble walling
> F30 Accessories for stone walling

**Carpentry and joinery**

> G20 Carpentry/timber framing/first fixing
> G32 Edge-supported woodwool decking
> K20 Timber board flooring
> L10 Timber windows
> L20 Timber doors
> L30 Timber stairs
> N11 Domestic fittings
> Y50 Thermal nsulation
> P21 Ironmongery

**Metalwork**

> L31 Balustrades

## Roofing

H60 Clay/concrete roof tiling
H61 Fibre-cement slating
H62 Natural slating
H63 Reconstructed stone slating
H71 Lead sheet coverings
H73 Copper sheet coverings
J41 Built-up roofing

## Asphalt work

J20 Mastic asphalt tanking
M11 Mastic asphalt flooring
J21 Mastic asphalt roofing

## Floor, wall and ceiling finishes

M10 Sand cement/granolithic screeds
M20 Plastered coatings
M30 Metal mesh lathing
K10 Plasterboard dry lining
M40 Quarry/ceramic tilling
M50 Plastic/lino tilling

## Plumbing and heating

N13 Sanitary fittings
R10 Rainwater pipes and gutters
R11 Foul drainage above ground
S10 Piped supply systems
Y21 Water tanks/ cisterns
Y23 Storage cylinders
S41 Fuel oil storage
T31 Low temperature hot water heating

## Glazing

L40 General glazing

## Painting and wallpapering

M60 Painting/clear finishing
M52 Decorative papers

## External works

R12 Drainage below ground
Q40 Fencing
Q10 Concrete kerbs and edgings
Q20 Hardcore/granular sub-bases
Q25  Slab/brick pavings

## Alterations and repairs

C10 Demolishing structures
C20 Spot items
C30 Shoring
D50 Underpinning

# Part One

## UNIT RATES

Demolition, excavation and filling

Concrete work

Brickwork and blockwork

Masonry

Carpentry and joinery

Metalwork

Roofing

Asphalt work

Floor, wall and ceiling finishings

Plumbing and heating

Glazing

Painting and wallpapering

External works

Alterations and repairs

Spot items

| | Unit | Labour hours | Labour cost £ | Plant £ | O & P £ | Total £ |
|---|---|---|---|---|---|---|

## DEMOLITION, EXCAVATION AND FILLING

### Demolition

Demolish ground level existing brick buildings

| | Unit | Labour hours | Labour cost £ | Plant £ | O & P £ | Total £ |
|---|---|---|---|---|---|---|
| single storey, detached | m3 | 1.60 | 22.40 | 0.00 | 3.36 | 25.76 |
| single storey, attached | m3 | 1.70 | 23.80 | 0.00 | 3.57 | 27.37 |
| two storey, attached | m3 | 1.80 | 25.20 | 0.00 | 3.78 | 28.98 |

Break up debris and arisings, load into skips or lorries

unreinforced concrete slabs

| | Unit | Labour hours | Labour cost £ | Plant £ | O & P £ | Total £ |
|---|---|---|---|---|---|---|
| 100mm thick | m2 | 0.80 | 11.20 | 0.65 | 1.78 | 13.63 |
| 150mm thick | m2 | 1.00 | 14.00 | 1.00 | 2.25 | 17.25 |
| 200mm thick | m2 | 1.40 | 19.60 | 1.30 | 3.14 | 24.04 |

reinforced concrete slabs

| | Unit | Labour hours | Labour cost £ | Plant £ | O & P £ | Total £ |
|---|---|---|---|---|---|---|
| 100mm thick | m2 | 1.40 | 19.60 | 1.30 | 3.14 | 24.04 |
| 150mm thick | m2 | 2.00 | 28.00 | 2.00 | 4.50 | 34.50 |
| 200mm thick | m2 | 2.60 | 36.40 | 2.65 | 5.86 | 44.91 |

reinforced suspended concrete slabs

| | Unit | Labour hours | Labour cost £ | Plant £ | O & P £ | Total £ |
|---|---|---|---|---|---|---|
| 100mm thick | m2 | 2.80 | 39.20 | 1.30 | 6.08 | 46.58 |
| 150mm thick | m2 | 3.20 | 44.80 | 2.00 | 7.02 | 53.82 |
| 200mm thick | m2 | 3.40 | 47.60 | 2.65 | 7.54 | 57.79 |

reinforced concrete beams

| | Unit | Labour hours | Labour cost £ | Plant £ | O & P £ | Total £ |
|---|---|---|---|---|---|---|
| 200 × 300mm | m | 1.25 | 17.50 | 0.27 | 2.67 | 20.44 |
| 250 × 350mm | m | 1.50 | 21.00 | 0.40 | 3.21 | 24.61 |
| 300 × 400mm | m | 1.75 | 24.50 | 0.52 | 3.75 | 28.77 |

reinforced concrete columns

| | Unit | Labour hours | Labour cost £ | Plant £ | O & P £ | Total £ |
|---|---|---|---|---|---|---|
| 200 × 300mm | m | 1.30 | 18.20 | 0.27 | 2.77 | 21.24 |
| 250 × 350mm | m | 1.60 | 22.40 | 0.40 | 3.42 | 26.22 |
| 300 × 400mm | m | 1.90 | 26.60 | 0.52 | 4.07 | 31.19 |

## 4 Unit rates

| | Unit | Labour hours | Labour cost £ | Plant £ | O & P £ | Total £ |
|---|---|---|---|---|---|---|
| **Demolish brick walls in cement mortar** | | | | | | |
| half brick thick | m2 | 0.80 | 11.20 | 0.00 | 1.68 | 12.88 |
| one brick thick | m2 | 1.40 | 19.60 | 0.00 | 2.94 | 22.54 |
| one and a half brick thic | m2 | 1.90 | 26.60 | 0.00 | 3.99 | 30.59 |
| two brick thick | m2 | 2.70 | 37.80 | 0.00 | 5.67 | 43.47 |
| **Demolish brick walls in cement lime mortar** | | | | | | |
| half brick thick | m2 | 0.70 | 9.80 | 0.00 | 1.47 | 11.27 |
| one brick thick | m2 | 1.30 | 18.20 | 0.00 | 2.73 | 20.93 |
| one and a half brick thic | m2 | 1.80 | 25.20 | 0.00 | 3.78 | 28.98 |
| two brick thick | m2 | 2.50 | 35.00 | 0.00 | 5.25 | 40.25 |
| **Carefully take down brick walls in cement lime mortar, clean bricks and lay aside for re-use** | | | | | | |
| half brick thick | m2 | 1.00 | 14.00 | 0.00 | 2.10 | 16.10 |
| one brick thick | m2 | 1.70 | 23.80 | 0.00 | 3.57 | 27.37 |
| one and a half brick thic | m2 | 2.20 | 30.80 | 0.00 | 4.62 | 35.42 |
| two brick thick | m2 | 3.20 | 44.80 | 0.00 | 6.72 | 51.52 |
| **Take out fireplace and surround, load into skips or lorries** | | | | | | |
| 1500 × 1200mm high | nr | 1.50 | 21.00 | 0.00 | 3.15 | 24.15 |
| 1500 × 1500mm high | nr | 1.75 | 24.50 | 0.00 | 3.68 | 28.18 |
| 1800 × 1500mm high | nr | 2.00 | 28.00 | 0.00 | 4.20 | 32.20 |
| **Demolish blockwork partitions** | | | | | | |
| 75mm thick | m2 | 0.80 | 11.20 | 0.00 | 1.68 | 12.88 |
| 100mm thick | m2 | 0.90 | 12.60 | 0.00 | 1.89 | 14.49 |
| 115mm thick | m2 | 1.00 | 14.00 | 0.00 | 2.10 | 16.10 |
| 140mm thick | m2 | 1.10 | 15.40 | 0.00 | 2.31 | 17.71 |
| 190mm thick | m2 | 1.20 | 16.80 | 0.00 | 2.52 | 19.32 |
| 215mm thick | m2 | 1.40 | 19.60 | 0.00 | 2.94 | 22.54 |

| | Unit | Labour hours | Labour cost £ | Plant £ | O & P £ | Total £ |
|---|---|---|---|---|---|---|
| **Excavation** | | | | | | |
| Excavate by hand | | | | | | |
| topsoil | | | | | | |
| 150mm thick | m2 | 0.35 | 4.90 | 0.00 | 0.74 | 5.64 |
| 200mm thick | m2 | 0.45 | 6.30 | 0.00 | 0.95 | 7.25 |
| 250mm thick | m2 | 0.60 | 8.40 | 0.00 | 1.26 | 9.66 |
| to reduce levels, depth not exceeding | | | | | | |
| 0.75m | m3 | 2.20 | 30.80 | 0.00 | 4.62 | 35.42 |
| 1.00m | m3 | 2.40 | 33.60 | 0.00 | 5.04 | 38.64 |
| 1.50m | m3 | 2.80 | 39.20 | 0.00 | 5.88 | 45.08 |
| 2.00m | m3 | 3.20 | 44.80 | 0.00 | 6.72 | 51.52 |
| to trenches 450mm wide, depth not exceeding | | | | | | |
| 0.75m | m3 | 2.50 | 35.00 | 0.00 | 5.25 | 40.25 |
| 1.00m | m3 | 2.70 | 37.80 | 0.00 | 5.67 | 43.47 |
| 1.50m | m3 | 3.00 | 42.00 | 0.00 | 6.30 | 48.30 |
| 2.00m | m3 | 3.40 | 47.60 | 0.00 | 7.14 | 54.74 |
| to pits, depth not exceeding | | | | | | |
| 0.75m | m3 | 2.50 | 35.00 | 0.00 | 5.25 | 40.25 |
| 1.00m | m3 | 2.70 | 37.80 | 0.00 | 5.67 | 43.47 |
| 1.50m | m3 | 3.10 | 43.40 | 0.00 | 6.51 | 49.91 |
| 2.00m | m3 | 3.50 | 49.00 | 0.00 | 7.35 | 56.35 |
| Extra for excavating through | | | | | | |
| rock | m3 | 10.00 | 140.00 | 0.00 | 21.00 | 161.00 |
| concrete | m3 | 8.00 | 112.00 | 0.00 | 16.80 | 128.80 |
| reinforced concrete | m3 | 9.00 | 126.00 | 0.00 | 18.90 | 144.90 |
| brickwork | m3 | 6.00 | 84.00 | 0.00 | 12.60 | 96.60 |

## 6 Unit rates

| | Unit | Labour hours | Labour cost £ | Plant £ | O & P £ | Total £ |
|---|---|---|---|---|---|---|
| **Excavate by machine** | | | | | | |
| | | | | | | |
| topsoil | | | | | | |
| 150mm thick | m2 | 0.04 | 0.56 | 1.44 | 0.30 | 2.30 |
| 200mm thick | m2 | 0.05 | 0.70 | 2.12 | 0.42 | 3.24 |
| 250mm thick | m2 | 0.06 | 0.84 | 2.88 | 0.56 | 4.28 |
| | | | | | | |
| to reduce levels, depth not exceeding | | | | | | |
| 0.75m | m3 | 0.10 | 1.40 | 1.75 | 0.47 | 3.62 |
| 1.00m | m3 | 0.12 | 1.68 | 1.54 | 0.48 | 3.70 |
| 1.50m | m3 | 0.14 | 1.96 | 1.75 | 0.56 | 4.27 |
| 2.00m | m3 | 0.16 | 2.24 | 2.00 | 0.64 | 4.88 |
| | | | | | | |
| to trenches 450mm wide, depth not exceeding | | | | | | |
| 0.75m | m3 | 0.35 | 4.90 | 4.80 | 1.46 | 11.16 |
| 1.00m | m3 | 0.40 | 5.60 | 4.40 | 1.50 | 11.50 |
| 1.50m | m3 | 0.45 | 6.30 | 4.80 | 1.67 | 12.77 |
| 2.00m | m3 | 0.50 | 7.00 | 5.10 | 1.82 | 13.92 |
| | | | | | | |
| to pits, depth not exceeding | | | | | | |
| 0.75m | m3 | 0.45 | 6.30 | 5.05 | 1.70 | 13.05 |
| 1.00m | m3 | 0.50 | 7.00 | 4.60 | 1.74 | 13.34 |
| 1.50m | m3 | 0.55 | 7.70 | 5.05 | 1.91 | 14.66 |
| 2.00m | m3 | 0.60 | 8.40 | 5.25 | 2.05 | 15.70 |
| | | | | | | |
| **Extra for excavating through** | | | | | | |
| | | | | | | |
| rock | m3 | 5.00 | 70.00 | 20.00 | 13.50 | 103.50 |
| concrete | m3 | 3.75 | 52.50 | 15.30 | 10.17 | 77.97 |
| reinforced concrete | m3 | 4.00 | 56.00 | 17.50 | 11.03 | 84.53 |
| brickwork | m3 | 3.80 | 53.20 | 14.20 | 10.11 | 77.51 |

| | Unit | Labour hours | Labour cost £ | Materials £ | O & P £ | Total £ |
|---|---|---|---|---|---|---|

**Earthwork support**

Earthwork support not exceeding
2m between opposite faces,
depth not exceeding 1m

| | Unit | Labour hours | Labour cost £ | Materials £ | O & P £ | Total £ |
|---|---|---|---|---|---|---|
| firm ground | m2 | 0.15 | 2.10 | 1.48 | 0.54 | 4.12 |
| loose ground | m2 | 0.85 | 11.90 | 2.98 | 2.23 | 17.11 |
| sand | m2 | 1.10 | 15.40 | 3.44 | 2.83 | 21.67 |

Earthwork support not exceeding
2m between opposite faces,
depth not exceeding 2m

| | Unit | Labour hours | Labour cost £ | Materials £ | O & P £ | Total £ |
|---|---|---|---|---|---|---|
| firm ground | m2 | 0.18 | 2.52 | 1.48 | 0.60 | 4.60 |
| loose ground | m2 | 0.90 | 12.60 | 2.98 | 2.34 | 17.92 |
| sand | m2 | 1.20 | 16.80 | 3.44 | 3.04 | 23.28 |

Earthwork support not exceeding
2m between opposite faces,
depth not exceeding 4m

| | Unit | Labour hours | Labour cost £ | Materials £ | O & P £ | Total £ |
|---|---|---|---|---|---|---|
| firm ground | m2 | 0.22 | 3.08 | 1.48 | 0.68 | 5.24 |
| loose ground | m2 | 0.95 | 13.30 | 2.98 | 2.44 | 18.72 |
| sand | m2 | 1.30 | 18.20 | 3.44 | 3.25 | 24.89 |

**Disposal by hand**

Load surplus excavated material
into barrows, wheel and deposit
in temporary spoil heaps, average
distance

| | Unit | Labour hours | Labour cost £ | Materials £ | O & P £ | Total £ |
|---|---|---|---|---|---|---|
| 15m | m3 | 1.25 | 17.50 | 0.00 | 2.63 | 20.13 |
| 25m | m3 | 1.45 | 20.30 | 0.00 | 3.05 | 23.35 |
| 50m | m3 | 1.70 | 23.80 | 0.00 | 3.57 | 27.37 |

| | Unit | Labour hours | Labour cost £ | Plant £ | O & P £ | Total £ |
|---|---|---|---|---|---|---|
| **Load surplus excavated material into barrows, wheel and spread over site, average distance** | | | | | | |
| 15m | m3 | 1.65 | 23.10 | 0.00 | 3.47 | 26.57 |
| 25m | m3 | 1.85 | 25.90 | 0.00 | 3.89 | 29.79 |
| 50m | m3 | 2.00 | 28.00 | 0.00 | 4.20 | 32.20 |
| **Load surplus excavated material into barrows, wheel and deposit in skips or lorries, average distance** | | | | | | |
| 15m | m3 | 1.20 | 16.80 | 0.00 | 2.52 | 19.32 |
| 25m | m3 | 1.40 | 19.60 | 0.00 | 2.94 | 22.54 |
| 50m | m3 | 1.65 | 23.10 | 0.00 | 3.47 | 26.57 |

**Disposal by machine**

| | Unit | Labour hours | Labour cost £ | Plant £ | O & P £ | Total £ |
|---|---|---|---|---|---|---|
| **Load and deposit in temporary spoil heaps, average distance** | | | | | | |
| 15m | m3 | 0.10 | 1.40 | 1.10 | 0.38 | 2.88 |
| 25m | m3 | 0.15 | 2.10 | 1.20 | 0.50 | 3.80 |
| 50m | m3 | 0.25 | 3.50 | 1.96 | 0.82 | 6.28 |
| **Load and spread over site, average distance** | | | | | | |
| 15m | m3 | 0.15 | 2.10 | 1.32 | 0.51 | 3.93 |
| 25m | m3 | 0.18 | 2.52 | 1.48 | 0.60 | 4.60 |
| 50m | m3 | 0.24 | 3.36 | 1.65 | 0.75 | 5.76 |
| **Load and deposit in skips or lorries, average distance** | | | | | | |
| 15m | m3 | 0.08 | 1.12 | 1.10 | 0.33 | 2.55 |
| 25m | m3 | 0.12 | 1.68 | 1.20 | 0.43 | 3.31 |
| 50m | m3 | 0.22 | 3.08 | 1.96 | 0.76 | 5.80 |

| | Unit | Labour hours | Labour cost £ | Materials £ | O & P £ | Total £ |
|---|---|---|---|---|---|---|

**Filling by hand**

Surplus excavated material
deposited and compacting in
layers

| | | | | | | |
|---|---|---|---|---|---|---|
| over 250mm thick | m3 | 1.20 | 16.80 | 0.00 | 2.52 | 19.32 |
| 100mm thick | m2 | 0.20 | 2.80 | 0.00 | 0.42 | 3.22 |
| 150mm thick | m2 | 0.35 | 4.90 | 0.00 | 0.74 | 5.64 |
| 200mm thick | m2 | 0.50 | 7.00 | 0.00 | 1.05 | 8.05 |

Imported sand deposited and
compacting in layers

| | | | | | | |
|---|---|---|---|---|---|---|
| over 250mm thick | m3 | 1.20 | 16.80 | 34.00 | 7.62 | 58.42 |
| 100mm thick | m2 | 0.20 | 2.80 | 3.40 | 0.93 | 7.13 |
| 150mm thick | m2 | 0.35 | 4.90 | 5.10 | 1.50 | 11.50 |
| 200mm thick | m2 | 0.50 | 7.00 | 6.80 | 2.07 | 15.87 |

Imported hardcore deposited
and compacting in layers

| | | | | | | |
|---|---|---|---|---|---|---|
| over 250mm thick | m3 | 1.20 | 16.80 | 21.02 | 5.67 | 43.49 |
| 100mm thick | m2 | 0.20 | 2.80 | 2.10 | 0.74 | 5.64 |
| 150mm thick | m2 | 0.35 | 4.90 | 3.15 | 1.21 | 9.26 |
| 200mm thick | m2 | 0.50 | 7.00 | 4.20 | 1.68 | 12.88 |

**Filling by machine**

Surplus excavated material
deposited and compacting in
layers

| | | | | | | |
|---|---|---|---|---|---|---|
| over 250mm thick | m3 | 0.40 | 5.60 | 0.00 | 0.84 | 6.44 |
| 100mm thick | m2 | 0.06 | 0.84 | 0.00 | 0.13 | 0.97 |
| 150mm thick | m2 | 0.08 | 1.12 | 0.00 | 0.17 | 1.29 |
| 200mm thick | m2 | 0.10 | 1.40 | 0.00 | 0.21 | 1.61 |

|  | Unit | Labour hours | Labour cost £ | Materials £ | O & P £ | Total £ |
|---|---|---|---|---|---|---|
| **Imported sand deposited and compacting in layers** | | | | | | |
| over 250mm thick | m3 | 0.40 | 5.60 | 34.00 | 5.94 | 45.54 |
| 100mm thick | m2 | 0.06 | 0.84 | 3.40 | 0.64 | 4.88 |
| 150mm thick | m2 | 0.08 | 1.12 | 5.10 | 0.93 | 7.15 |
| 200mm thick | m2 | 0.10 | 1.40 | 6.80 | 1.23 | 9.43 |
| **Imported hardcore deposited and compacting in layers** | | | | | | |
| over 250mm thick | m3 | 0.40 | 5.60 | 21.02 | 3.99 | 30.61 |
| 100mm thick | m2 | 0.06 | 0.84 | 2.10 | 0.44 | 3.38 |
| 150mm thick | m2 | 0.08 | 1.12 | 3.15 | 0.64 | 4.91 |
| 200mm thick | m2 | 0.10 | 1.40 | 4.24 | 0.85 | 6.49 |
| **Surface treatments** | | | | | | |
| Level and compact bottom of excavation with vibrating roller | m2 | 0.10 | 1.40 | 0.00 | 0.21 | 1.61 |
| Blind filling surfaces with sand 50mm thick | m2 | 0.12 | 1.68 | 1.62 | 0.50 | 3.80 |

|  | Unit | Labour hours | Labour cost £ | Materials £ | O & P £ | Total £ |
|---|---|---|---|---|---|---|
| **CONCRETE WORK** | | | | | | |
| **Ready-mixed concrete 1:3:6** **40mm aggregate** | | | | | | |
| Foundations in trenches, thickness | | | | | | |
| 150 to 300mm | m3 | 1.85 | 25.90 | 85.62 | 16.73 | 128.25 |
| 300 to 450mm | m3 | 1.35 | 18.90 | 85.62 | 15.68 | 120.20 |
| over 450mm | m3 | 1.15 | 16.10 | 85.62 | 15.26 | 116.98 |
| Isolated column bases | | | | | | |
| 150 to 300mm | m3 | 2.00 | 28.00 | 85.62 | 17.04 | 130.66 |
| 300 to 450mm | m3 | 1.50 | 21.00 | 85.62 | 15.99 | 122.61 |
| over 450mm | m3 | 1.30 | 18.20 | 85.62 | 15.57 | 119.39 |
| Filling to cavity walls | m3 | 3.95 | 55.30 | 85.62 | 21.14 | 162.06 |
| **Ready-mixed concrete 1:2:4** **20mm aggregate** | | | | | | |
| Beds | | | | | | |
| 150 to 300mm | m3 | 2.40 | 33.60 | 91.44 | 18.76 | 143.80 |
| 300 to 450mm | m3 | 2.10 | 29.40 | 91.44 | 18.13 | 138.97 |
| over 450mm | m3 | 1.80 | 25.20 | 91.44 | 17.50 | 134.14 |
| Slabs | | | | | | |
| 150 to 300mm | m3 | 3.20 | 44.80 | 91.44 | 20.44 | 156.68 |
| 300 to 450mm | m3 | 2.85 | 39.90 | 91.44 | 19.70 | 151.04 |
| Walls | | | | | | |
| 150 to 300mm | m3 | 4.25 | 59.50 | 91.44 | 22.64 | 173.58 |
| 300 to 450mm | m3 | 3.95 | 55.30 | 91.44 | 22.01 | 168.75 |

| | Unit | Labour hours | Labour cost £ | Materials £ | O & P £ | Total £ |
|---|---|---|---|---|---|---|
| **Casings** | | | | | | |
| isolated beams | m3 | 5.00 | 70.00 | 91.44 | 24.22 | 185.66 |
| isolated deep beams | m3 | 4.50 | 63.00 | 91.44 | 23.17 | 177.61 |
| attached deep beams | m3 | 5.00 | 70.00 | 91.44 | 24.22 | 185.66 |
| columns | m3 | 5.00 | 70.00 | 91.44 | 24.22 | 185.66 |
| staircases | m3 | 6.00 | 84.00 | 91.44 | 26.32 | 201.76 |
| Upstands and kerbs | m3 | 4.40 | 61.60 | 91.44 | 22.96 | 176.00 |

**Site-mixed concrete 1:3:6
40mm aggregate**

Foundations in trenches, thickness

| | Unit | Labour hours | Labour cost £ | Materials £ | O & P £ | Total £ |
|---|---|---|---|---|---|---|
| 150 to 300mm | m3 | 2.85 | 39.90 | 78.33 | 17.73 | 135.96 |
| 300 to 450mm | m3 | 2.35 | 32.90 | 78.33 | 16.68 | 127.91 |
| over 450mm | m3 | 2.15 | 30.10 | 78.33 | 16.26 | 124.69 |

Isolated column bases

| | Unit | Labour hours | Labour cost £ | Materials £ | O & P £ | Total £ |
|---|---|---|---|---|---|---|
| 150 to 300mm | m3 | 3.00 | 42.00 | 78.33 | 18.05 | 138.38 |
| 300 to 450mm | m3 | 2.50 | 35.00 | 78.33 | 17.00 | 130.33 |
| over 450mm | m3 | 2.30 | 32.20 | 78.33 | 16.58 | 127.11 |
| Filling to cavity walls | m3 | 4.95 | 69.30 | 78.33 | 22.14 | 169.77 |

**Site-mixed concrete 1:2:4
20mm aggregate**

Beds

| | Unit | Labour hours | Labour cost £ | Materials £ | O & P £ | Total £ |
|---|---|---|---|---|---|---|
| 150 to 300mm | m3 | 3.40 | 47.60 | 87.56 | 20.27 | 155.43 |
| 300 to 450mm | m3 | 3.10 | 43.40 | 87.56 | 19.64 | 150.60 |
| over 450mm | m3 | 1.90 | 26.60 | 87.56 | 17.12 | 131.28 |

Slabs

| | Unit | Labour hours | Labour cost £ | Materials £ | O & P £ | Total £ |
|---|---|---|---|---|---|---|
| 150 to 300mm | m3 | 4.20 | 58.80 | 87.56 | 21.95 | 168.31 |
| 300 to 450mm | m3 | 3.85 | 53.90 | 87.56 | 21.22 | 162.68 |

| | Unit | Labour hours | Labour cost £ | Materials £ | O & P £ | Total £ |
|---|---|---|---|---|---|---|
| **Walls** | | | | | | |
| 150 to 300mm | m3 | 5.25 | 73.50 | 87.56 | 24.16 | 185.22 |
| 300 to 450mm | m3 | 4.95 | 69.30 | 87.56 | 23.53 | 180.39 |
| **Casings** | | | | | | |
| isolated beams | m3 | 6.00 | 84.00 | 87.56 | 25.73 | 197.29 |
| isolated deep beams | m3 | 5.50 | 77.00 | 87.56 | 24.68 | 189.24 |
| attached deep beams | m3 | 6.00 | 84.00 | 87.56 | 25.73 | 197.29 |
| columns | m3 | 6.00 | 84.00 | 87.56 | 25.73 | 197.29 |
| staircases | m3 | 7.00 | 98.00 | 87.56 | 27.83 | 213.39 |
| **Upstands and kerbs** | m3 | 5.50 | 77.00 | 87.56 | 24.68 | 189.24 |
| **Formwork** | | | | | | |
| **Plain vertical to sides of foundations** | | | | | | |
| over 1m | m2 | 2.30 | 32.20 | 9.87 | 6.31 | 48.38 |
| not exceeding 250mm | m | 0.75 | 10.50 | 3.46 | 2.09 | 16.05 |
| 250 to 500mm | m | 1.25 | 17.50 | 5.98 | 3.52 | 27.00 |
| 500mm to 1m | m | 1.70 | 23.80 | 9.87 | 5.05 | 38.72 |
| **Plain vertical to sides of foundations, left in** | | | | | | |
| over 1m | m2 | 2.15 | 30.10 | 24.04 | 8.12 | 62.26 |
| not exceeding 250mm | m | 0.65 | 9.10 | 8.32 | 2.61 | 20.03 |
| 250 to 500mm | m | 1.15 | 16.10 | 15.76 | 4.78 | 36.64 |
| 500mm to 1m | m | 1.60 | 22.40 | 24.00 | 6.96 | 53.36 |
| **Plain vertical to sides of ground beams and beds** | | | | | | |
| over 1m | m2 | 2.40 | 33.60 | 9.87 | 6.52 | 49.99 |
| not exceeding 250mm | m | 0.82 | 11.48 | 3.46 | 2.24 | 17.18 |
| 250 to 500mm | m | 1.34 | 18.76 | 5.98 | 3.71 | 28.45 |
| 500mm to 1m | m | 1.78 | 24.92 | 9.87 | 5.22 | 40.01 |

| | Unit | Labour hours | Labour cost £ | Materials £ | O & P £ | Total £ |
|---|---|---|---|---|---|---|
| **Plain vertical to sides of ground beams and beds, left in** | | | | | | |
| over 1m | m2 | 2.25 | 31.50 | 24.04 | 8.33 | 63.87 |
| not exceeding 250mm | m | 0.75 | 10.50 | 8.32 | 2.82 | 21.64 |
| 250 to 500mm | m | 1.25 | 17.50 | 15.76 | 4.99 | 38.25 |
| 500mm to 1m | m | 1.70 | 23.80 | 24.00 | 7.17 | 54.97 |
| **Edges of suspended slabs** | | | | | | |
| not exceeding 250mm | m | 0.95 | 13.30 | 3.46 | 2.51 | 19.27 |
| 250 to 500mm | m | 1.40 | 19.60 | 5.98 | 3.84 | 29.42 |
| **Sides of upstands** | | | | | | |
| over 1m | m2 | 2.60 | 36.40 | 9.87 | 6.94 | 53.21 |
| not exceeding 250mm | m | 1.00 | 14.00 | 3.46 | 2.62 | 20.08 |
| 250 to 500mm | m | 1.50 | 21.00 | 5.98 | 4.05 | 31.03 |
| 500mm to 1m | m | 1.90 | 26.60 | 9.87 | 5.47 | 41.94 |
| **Steps in top surfaces** | | | | | | |
| not exceeding 250mm | m | 0.75 | 10.50 | 3.46 | 2.09 | 16.05 |
| 250 to 500mm | m | 1.25 | 17.50 | 5.98 | 3.52 | 27.00 |
| 500mm to 1m | m | 1.70 | 23.80 | 9.87 | 5.05 | 38.72 |
| **Steps in soffits** | | | | | | |
| not exceeding 250mm | m | 1.00 | 14.00 | 3.46 | 2.62 | 20.08 |
| 250 to 500mm | m | 1.50 | 21.00 | 5.98 | 4.05 | 31.03 |
| 500mm to 1m | m | 1.90 | 26.60 | 9.87 | 5.47 | 41.94 |
| **Machine bases and plinths** | | | | | | |
| over 1m | m2 | 2.30 | 32.20 | 9.87 | 6.31 | 48.38 |
| not exceeding 250mm | m | 0.75 | 10.50 | 3.46 | 2.09 | 16.05 |
| 250 to 500mm | m | 1.25 | 17.50 | 5.98 | 3.52 | 27.00 |
| 500mm to 1m | m | 1.70 | 23.80 | 9.87 | 5.05 | 38.72 |

|  | Unit | Labour hours | Labour cost £ | Materials £ | O & P £ | Total £ |
|---|---|---|---|---|---|---|
| **Horizontal to soffits of slabs, thickness** | | | | | | |
| **not exceeding 200mm** | | | | | | |
| height to soffit, not exceeding 1.5m | m2 | 2.20 | 30.80 | 9.87 | 6.10 | 46.77 |
| height to soffit, 1.5 to 3m | m2 | 2.10 | 29.40 | 9.87 | 5.89 | 45.16 |
| height to soffit, over 3m | m2 | 2.00 | 28.00 | 9.87 | 5.68 | 43.55 |
| **200 to 300mm** | | | | | | |
| height to soffit, not exceeding 1.5m | m2 | 2.40 | 33.60 | 9.87 | 6.52 | 49.99 |
| height to soffit, 1.5 to 3m | m2 | 2.30 | 32.20 | 9.87 | 6.31 | 48.38 |
| height to soffit, over 3m | m2 | 2.20 | 30.80 | 9.87 | 6.10 | 46.77 |
| **300 to 400mm** | | | | | | |
| height to soffit, not exceeding 1.5m | m2 | 2.60 | 36.40 | 9.87 | 6.94 | 53.21 |
| height to soffit, 1.5 to 3m | m2 | 2.50 | 35.00 | 9.87 | 6.73 | 51.60 |
| height to soffit, over 3m | m2 | 2.40 | 33.60 | 9.87 | 6.52 | 49.99 |
| **Horizontal to soffits of slabs, left in, thickness** | | | | | | |
| **not exceeding 200mm** | | | | | | |
| height to soffit, not exceeding 1.5m | m2 | 2.25 | 31.50 | 24.00 | 8.33 | 63.83 |
| height to soffit, 1.5 to 3m | m2 | 2.20 | 30.80 | 24.00 | 8.22 | 63.02 |
| height to soffit, over 3m | m2 | 2.10 | 29.40 | 24.00 | 8.01 | 61.41 |
| **200 to 300mm** | | | | | | |
| height to soffit, not exceeding 1.5m | m2 | 2.35 | 32.90 | 24.00 | 8.54 | 65.44 |
| height to soffit, 1.5 to 3m | m2 | 2.30 | 32.20 | 24.00 | 8.43 | 64.63 |
| height to soffit, over 3m | m2 | 2.20 | 30.80 | 24.00 | 8.22 | 63.02 |

| | Unit | Labour hours | Labour cost £ | Materials £ | O & P £ | Total £ |
|---|---|---|---|---|---|---|
| **Horizontal to soffits of slabs, left in, thickness (cont'd)** | | | | | | |
| 300 to 400mm | | | | | | |
| height to soffit, not exceeding 1.5m | m2 | 2.45 | 34.30 | 24.00 | 8.75 | 67.05 |
| height to soffit, 1.5 to 3m | m2 | 2.40 | 33.60 | 24.00 | 8.64 | 66.24 |
| height to soffit, over 3m | m2 | 2.30 | 32.20 | 24.00 | 8.43 | 64.63 |
| **Walls** | | | | | | |
| vertical | m2 | 2.40 | 33.60 | 9.87 | 6.52 | 49.99 |
| vertical interrupted | m2 | 2.50 | 35.00 | 9.87 | 6.73 | 51.60 |
| **Beams** | | | | | | |
| attached to slabs, height to soffit | | | | | | |
| 1.5 to 3m | m2 | 3.20 | 44.80 | 9.87 | 8.20 | 62.87 |
| 3 to 4.5m | m2 | 3.30 | 46.20 | 9.87 | 8.41 | 64.48 |
| over 4.5m | m2 | 3.50 | 49.00 | 9.87 | 8.83 | 67.70 |
| **Columns** | | | | | | |
| attached to walls, height to soffit | | | | | | |
| 1.5 to 3m | m2 | 3.20 | 44.80 | 9.87 | 8.20 | 62.87 |
| 3 to 4.5m | m2 | 3.30 | 46.20 | 9.87 | 8.41 | 64.48 |
| over 4.5m | m2 | 3.50 | 49.00 | 9.87 | 8.83 | 67.70 |
| isolated, height to soffit | | | | | | |
| 1.5 to 3m | m2 | 3.40 | 47.60 | 9.87 | 8.62 | 66.09 |
| 3 to 4.5m | m2 | 3.50 | 49.00 | 9.87 | 8.83 | 67.70 |
| over 4.5m | m2 | 3.60 | 50.40 | 9.87 | 9.04 | 69.31 |
| **Mortice in concrete for rag bolt, grout in cement mortar (1:3), depth** | | | | | | |
| 50mm | nr | 0.30 | 4.20 | 1.02 | 0.78 | 6.00 |
| 100mm | nr | 0.35 | 4.90 | 2.15 | 1.06 | 8.11 |
| 150mm | nr | 0.40 | 5.60 | 3.50 | 1.37 | 10.47 |

|  | Unit | Labour hours | Labour cost £ | Materials £ | O & P £ | Total £ |
|---|---|---|---|---|---|---|
| **Reinforcement** | | | | | | |
| Plain round steel reinforcement bars, straight or bent | | | | | | |
| 6mm diameter | m | 0.02 | 0.28 | 0.21 | 0.07 | 0.56 |
| 8mm diameter | m | 0.03 | 0.42 | 0.28 | 0.11 | 0.81 |
| 10mm diameter | m | 0.04 | 0.56 | 0.36 | 0.14 | 1.06 |
| 12mm diameter | m | 0.05 | 0.70 | 0.48 | 0.18 | 1.36 |
| 16mm diameter | m | 0.06 | 0.84 | 0.74 | 0.24 | 1.82 |
| 20mm diameter | m | 0.07 | 0.98 | 1.15 | 0.32 | 2.45 |
| 25mm diameter | m | 0.08 | 1.12 | 1.80 | 0.44 | 3.36 |
| High yield deformed steel reinforcement bars, straight or bent | | | | | | |
| 8mm diameter | m | 0.03 | 0.42 | 0.26 | 0.10 | 0.78 |
| 10mm diameter | m | 0.04 | 0.56 | 0.36 | 0.14 | 1.06 |
| 12mm diameter | m | 0.05 | 0.70 | 0.43 | 0.17 | 1.30 |
| 16mm diameter | m | 0.06 | 0.84 | 0.52 | 0.20 | 1.56 |
| 20mm diameter | m | 0.07 | 0.98 | 1.20 | 0.33 | 2.51 |
| 25mm diameter | m | 0.08 | 1.12 | 1.85 | 0.45 | 3.42 |
| Steel fabric reinforcement laid in concrete beds | | | | | | |
| A98 weighing 1.54kg per m2 | m2 | 0.12 | 1.68 | 1.37 | 0.46 | 3.51 |
| A142 weighing 2.22kg per m2 | m2 | 0.14 | 1.96 | 1.42 | 0.51 | 3.89 |
| A193 weighing 3.02kg per m2 | m2 | 0.17 | 2.38 | 1.92 | 0.65 | 4.95 |
| A252 weighing 3.95kg per m2 | m2 | 0.19 | 2.66 | 2.58 | 0.79 | 6.03 |
| B196 weighing 3.05kg per m2 | m2 | 0.17 | 2.38 | 2.16 | 0.68 | 5.22 |
| B283 weighing 3.73kg per m2 | m2 | 0.18 | 2.52 | 2.46 | 0.75 | 5.73 |
| B385 weighing 4.53kg per m2 | m2 | 0.20 | 2.80 | 2.97 | 0.87 | 6.64 |
| B503 weighing 5.93kg per m2 | m2 | 0.22 | 3.08 | 3.78 | 1.03 | 7.89 |
| B785 weighing 8.14kg per m2 | m2 | 0.24 | 3.36 | 5.33 | 1.30 | 9.99 |
| C283 weighing 2.61kg per m2 | m2 | 0.15 | 2.10 | 1.61 | 0.56 | 4.27 |
| C385 weighing 3.41kg per m2 | m2 | 0.18 | 2.52 | 2.29 | 0.72 | 5.53 |
| C503 weighing 4.34kg per m2 | m2 | 0.20 | 2.80 | 3.78 | 0.99 | 7.57 |
| C636 weighing 5.55kg per m2 | m2 | 0.21 | 2.94 | 3.37 | 0.95 | 7.26 |
| C785 weighing 6.72kg per m2 | m2 | 0.23 | 3.22 | 5.33 | 1.28 | 9.83 |

| | Unit | Labour hours | Labour cost £ | Materials £ | O & P £ | Total £ |
|---|---|---|---|---|---|---|

## Joints

Impregnated fibreboard expansion
joint, width

| | | | | | | |
|---|---|---|---|---|---|---|
| 12.5mm thick | | | | | | |
| not exceeding 150mm thick | m | 0.14 | 1.96 | 1.70 | 0.55 | 4.21 |
| 150 to 300mm thick | m | 0.20 | 2.80 | 2.83 | 0.84 | 6.47 |
| 300 to 450mm thick | m | 0.22 | 3.08 | 4.09 | 1.08 | 8.25 |
| 20mm thick | | | | | | |
| not exceeding 150mm thick | m | 0.16 | 2.24 | 2.50 | 0.71 | 5.45 |
| 150 to 300mm thick | m | 0.22 | 3.08 | 3.76 | 1.03 | 7.87 |
| 300 to 450mm thick | m | 0.24 | 3.36 | 5.63 | 1.35 | 10.34 |
| 25mm thick | | | | | | |
| not exceeding 150mm thick | m | 0.18 | 2.52 | 2.76 | 0.79 | 6.07 |
| 150 to 300mm thick | m | 0.24 | 3.36 | 4.31 | 1.15 | 8.82 |
| 300 to 450mm thick | m | 0.26 | 3.64 | 6.05 | 1.45 | 11.14 |

## Concrete finishes

Treat surfaces of unset concrete

| | | | | | | |
|---|---|---|---|---|---|---|
| mechanical tamping | m2 | 0.07 | 0.98 | 0.00 | 0.15 | 1.13 |
| power floating | m2 | 0.18 | 2.52 | 0.00 | 0.38 | 2.90 |
| trowelling | m2 | 0.24 | 3.36 | 0.00 | 0.50 | 3.86 |
| spade finish | m2 | 0.20 | 2.80 | 0.00 | 0.42 | 3.22 |
| wood float finish | m2 | 0.16 | 2.24 | 0.00 | 0.34 | 2.58 |

## Precast concrete

Reinforced concrete lintels
bedded in cement mortar (1:3)

| | | | | | | |
|---|---|---|---|---|---|---|
| 100 × 75 × 900mm | nr | 0.35 | 4.90 | 7.08 | 1.80 | 13.78 |
| 100 × 75 × 1000mm | nr | 0.40 | 5.60 | 7.96 | 2.03 | 15.59 |
| 100 × 75 × 1200mm | nr | 0.50 | 7.00 | 8.51 | 2.33 | 17.84 |
| 150 × 75 × 900mm | nr | 0.45 | 6.30 | 7.64 | 2.09 | 16.03 |
| 150 × 75 × 1000mm | nr | 0.45 | 6.30 | 8.64 | 2.24 | 17.18 |
| 150 × 75 × 1200mm | nr | 0.55 | 7.70 | 10.36 | 2.71 | 20.77 |

| | Unit | Labour hours | Labour cost £ | Materials £ | O & P £ | Total £ |
|---|---|---|---|---|---|---|
| Precast concrete copings bedded in cement mortar (1:3) | | | | | | |
| 75 × 150mm | m | 0.75 | 10.50 | 7.63 | 2.72 | 20.85 |
| 75 × 300mm | m | 0.90 | 12.60 | 14.96 | 4.13 | 31.69 |

| | Unit | Labour hours | Labour cost £ | Materials £ | O & P £ | Total £ |
|---|---|---|---|---|---|---|

## BRICKWORK AND BLOCKWORK

### Brickwork

Common bricks basic price £140
per thousand in cement mortar
(1:3)

walls

| | Unit | Labour hours | Labour cost £ | Materials £ | O & P £ | Total £ |
|---|---|---|---|---|---|---|
| half brick thick | m2 | 1.70 | 39.10 | 11.45 | 7.58 | 58.13 |
| half brick thick overhand | m2 | 2.20 | 50.60 | 11.45 | 9.31 | 71.36 |
| half brick thick against other work | m2 | 1.90 | 43.70 | 11.45 | 8.27 | 63.42 |
| half brick thick curved | m2 | 2.30 | 52.90 | 11.45 | 9.65 | 74.00 |
| one brick thick | m2 | 2.80 | 64.40 | 22.90 | 13.10 | 100.40 |
| one brick thick curved | m2 | 3.40 | 78.20 | 22.90 | 15.17 | 116.27 |
| one and a half brick thick | m2 | 3.50 | 80.50 | 34.35 | 17.23 | 132.08 |
| two brick thick | m2 | 4.20 | 96.60 | 45.80 | 21.36 | 163.76 |
| two brick thick battered | m2 | 4.80 | 110.40 | 45.80 | 23.43 | 179.63 |

walls, facework one side

| | Unit | Labour hours | Labour cost £ | Materials £ | O & P £ | Total £ |
|---|---|---|---|---|---|---|
| half brick thick | m2 | 1.80 | 41.40 | 11.45 | 7.93 | 60.78 |
| half brick thick overhand | m2 | 2.30 | 52.90 | 11.45 | 9.65 | 74.00 |
| half brick thick against other work | m2 | 2.00 | 46.00 | 11.45 | 8.62 | 66.07 |
| half brick thick curved | m2 | 2.40 | 55.20 | 11.45 | 10.00 | 76.65 |
| one brick thick | m2 | 2.90 | 66.70 | 22.90 | 13.44 | 103.04 |
| one brick thick curved | m2 | 3.50 | 80.50 | 22.90 | 15.51 | 118.91 |
| one and a half brick thick | m2 | 3.60 | 82.80 | 34.35 | 17.57 | 134.72 |
| two brick thick | m2 | 4.30 | 98.90 | 45.80 | 21.71 | 166.41 |
| two brick thick battered | m2 | 4.90 | 112.70 | 45.80 | 23.78 | 182.28 |

walls, facework both sides

| | Unit | Labour hours | Labour cost £ | Materials £ | O & P £ | Total £ |
|---|---|---|---|---|---|---|
| half brick thick | m2 | 1.90 | 43.70 | 11.45 | 8.27 | 63.42 |
| half brick thick overhand | m2 | 2.40 | 55.20 | 11.45 | 10.00 | 76.65 |
| half brick thick against other work | m2 | 2.10 | 48.30 | 11.45 | 8.96 | 68.71 |
| half brick thick curved | m2 | 2.50 | 57.50 | 11.45 | 10.34 | 79.29 |
| one brick thick | m2 | 3.00 | 69.00 | 22.90 | 13.79 | 105.69 |

| | Unit | Labour hours | Labour cost £ | Materials £ | O & P £ | Total £ |
|---|---|---|---|---|---|---|
| one brick thick curved | m2 | 3.60 | 82.80 | 22.90 | 15.86 | 121.56 |
| one and a half brick thick | m2 | 3.70 | 85.10 | 34.35 | 17.92 | 137.37 |
| two brick thick | m2 | 4.40 | 101.20 | 45.80 | 22.05 | 169.05 |
| two brick thick battered | m2 | 5.00 | 115.00 | 45.80 | 24.12 | 184.92 |

Common bricks basic price £200
per thousand in cement mortar
(1:3)

walls

| | Unit | Labour hours | Labour cost £ | Materials £ | O & P £ | Total £ |
|---|---|---|---|---|---|---|
| half brick thick | m2 | 1.70 | 39.10 | 16.60 | 8.36 | 64.06 |
| half brick thick overhand | m2 | 2.20 | 50.60 | 16.60 | 10.08 | 77.28 |
| half brick thick against other work | m2 | 1.90 | 43.70 | 16.60 | 9.05 | 69.35 |
| half brick thick curved | m2 | 2.30 | 52.90 | 16.60 | 10.43 | 79.93 |
| one brick thick | m2 | 2.80 | 64.40 | 33.20 | 14.64 | 112.24 |
| one brick thick curved | m2 | 3.40 | 78.20 | 33.20 | 16.71 | 128.11 |
| one and a half brick thick | m2 | 3.50 | 80.50 | 49.80 | 19.55 | 149.85 |
| two brick thick | m2 | 4.20 | 96.60 | 86.40 | 27.45 | 210.45 |
| two brick thick battered | m2 | 4.80 | 110.40 | 45.80 | 23.43 | 179.63 |

walls, facework one side

| | Unit | Labour hours | Labour cost £ | Materials £ | O & P £ | Total £ |
|---|---|---|---|---|---|---|
| half brick thick | m2 | 1.90 | 43.70 | 16.60 | 9.05 | 69.35 |
| half brick thick overhand | m2 | 2.40 | 55.20 | 16.60 | 10.77 | 82.57 |
| half brick thick against other work | m2 | 2.10 | 48.30 | 16.60 | 9.74 | 74.64 |
| half brick thick curved | m2 | 2.50 | 57.50 | 16.60 | 11.12 | 85.22 |
| one brick thick | m2 | 3.00 | 69.00 | 33.20 | 15.33 | 117.53 |
| one brick thick curved | m2 | 3.60 | 82.80 | 33.20 | 17.40 | 133.40 |
| one and a half brick thick | m2 | 3.70 | 85.10 | 49.80 | 20.24 | 155.14 |
| two brick thick | m2 | 4.40 | 101.20 | 86.40 | 28.14 | 215.74 |
| two brick thick battered | m2 | 5.00 | 115.00 | 86.40 | 30.21 | 231.61 |

walls, facework both sides

| | Unit | Labour hours | Labour cost £ | Materials £ | O & P £ | Total £ |
|---|---|---|---|---|---|---|
| half brick thick | m2 | 1.90 | 43.70 | 16.60 | 9.05 | 69.35 |
| half brick thick overhand | m2 | 2.40 | 55.20 | 16.60 | 10.77 | 82.57 |
| half brick thick against other work | m2 | 2.10 | 48.30 | 16.60 | 9.74 | 74.64 |
| half brick thick curved | m2 | 2.50 | 57.50 | 16.60 | 11.12 | 85.22 |
| one brick thick | m2 | 3.00 | 69.00 | 33.20 | 15.33 | 117.53 |

|  | Unit | Labour hours | Labour cost £ | Materials £ | O & P £ | Total £ |
|---|---|---|---|---|---|---|
| one brick thick curved | m2 | 3.60 | 82.80 | 33.20 | 17.40 | 133.40 |
| one and a half brick thick | m2 | 3.70 | 85.10 | 49.80 | 20.24 | 155.14 |
| two brick thick | m2 | 4.40 | 101.20 | 86.40 | 28.14 | 215.74 |
| two brick thick battered | m2 | 5.00 | 115.00 | 86.40 | 30.21 | 231.61 |

Common bricks basic price £140
per thousand in gauged mortar
(1:1:6)

walls

|  | Unit | Labour hours | Labour cost £ | Materials £ | O & P £ | Total £ |
|---|---|---|---|---|---|---|
| half brick thick | m2 | 1.70 | 39.10 | 11.30 | 7.56 | 57.96 |
| half brick thick overhand | m2 | 2.20 | 50.60 | 11.30 | 9.29 | 71.19 |
| half brick thick against other work | m2 | 1.90 | 43.70 | 11.30 | 8.25 | 63.25 |
| half brick thick curved | m2 | 2.30 | 52.90 | 11.30 | 9.63 | 73.83 |
| one brick thick | m2 | 2.80 | 64.40 | 22.60 | 13.05 | 100.05 |
| one brick thick curved | m2 | 3.40 | 78.20 | 22.60 | 15.12 | 115.92 |
| one and a half brick thick | m2 | 3.50 | 80.50 | 33.90 | 17.16 | 131.56 |
| two brick thick | m2 | 4.20 | 96.60 | 45.20 | 21.27 | 163.07 |
| two brick thick battered | m2 | 4.80 | 110.40 | 45.20 | 23.34 | 178.94 |

walls, facework one side

|  | Unit | Labour hours | Labour cost £ | Materials £ | O & P £ | Total £ |
|---|---|---|---|---|---|---|
| half brick thick | m2 | 1.80 | 41.40 | 11.30 | 7.91 | 60.61 |
| half brick thick overhand | m2 | 2.30 | 52.90 | 11.30 | 9.63 | 73.83 |
| half brick thick against other work | m2 | 2.00 | 46.00 | 11.30 | 8.60 | 65.90 |
| half brick thick curved | m2 | 2.40 | 55.20 | 11.30 | 9.98 | 76.48 |
| one brick thick | m2 | 2.90 | 66.70 | 22.60 | 13.40 | 102.70 |
| one brick thick curved | m2 | 3.50 | 80.50 | 22.60 | 15.47 | 118.57 |
| one and a half brick thick | m2 | 3.60 | 82.80 | 33.90 | 17.51 | 134.21 |
| two brick thick | m2 | 4.30 | 98.90 | 45.20 | 21.62 | 165.72 |
| two brick thick battered | m2 | 4.90 | 112.70 | 45.20 | 23.69 | 181.59 |

walls, facework both sides

|  | Unit | Labour hours | Labour cost £ | Materials £ | O & P £ | Total £ |
|---|---|---|---|---|---|---|
| half brick thick | m2 | 1.90 | 43.70 | 11.30 | 8.25 | 63.25 |
| half brick thick overhand | m2 | 2.40 | 55.20 | 11.30 | 9.98 | 76.48 |
| half brick thick against other work | m2 | 2.10 | 48.30 | 11.30 | 8.94 | 68.54 |
| half brick thick curved | m2 | 2.50 | 57.50 | 11.30 | 10.32 | 79.12 |
| one brick thick | m2 | 3.00 | 69.00 | 22.60 | 13.74 | 105.34 |

| | Unit | Labour hours | Labour cost £ | Materials £ | O & P £ | Total £ |
|---|---|---|---|---|---|---|
| one brick thick curved | m2 | 3.60 | 82.80 | 22.60 | 15.81 | 121.21 |
| one and a half brick thick | m2 | 3.70 | 85.10 | 33.90 | 17.85 | 136.85 |
| two brick thick | m2 | 4.40 | 101.20 | 45.20 | 21.96 | 168.36 |
| two brick thick battered | m2 | 5.00 | 115.00 | 45.20 | 24.03 | 184.23 |

Common bricks basic price £200
per thousand in gauged mortar
(1:1:6)

walls

| | Unit | Labour hours | Labour cost £ | Materials £ | O & P £ | Total £ |
|---|---|---|---|---|---|---|
| half brick thick | m2 | 1.70 | 39.10 | 16.45 | 8.33 | 63.88 |
| half brick thick overhand | m2 | 2.20 | 50.60 | 16.45 | 10.06 | 77.11 |
| half brick thick against other work | m2 | 1.90 | 43.70 | 16.45 | 9.02 | 69.17 |
| half brick thick curved | m2 | 2.30 | 52.90 | 16.45 | 10.40 | 79.75 |
| one brick thick | m2 | 2.80 | 64.40 | 32.90 | 14.60 | 111.90 |
| one brick thick curved | m2 | 3.40 | 78.20 | 32.90 | 16.67 | 127.77 |
| one and a half brick thick | m2 | 3.50 | 80.50 | 49.35 | 19.48 | 149.33 |
| two brick thick | m2 | 4.20 | 96.60 | 65.80 | 24.36 | 186.76 |
| two brick thick battered | m2 | 4.80 | 110.40 | 65.80 | 26.43 | 202.63 |

walls, facework one side

| | Unit | Labour hours | Labour cost £ | Materials £ | O & P £ | Total £ |
|---|---|---|---|---|---|---|
| half brick thick | m2 | 1.80 | 41.40 | 16.45 | 8.68 | 66.53 |
| half brick thick overhand | m2 | 2.30 | 52.90 | 16.45 | 10.40 | 79.75 |
| half brick thick against other work | m2 | 2.00 | 46.00 | 16.45 | 9.37 | 71.82 |
| half brick thick curved | m2 | 2.40 | 55.20 | 16.45 | 10.75 | 82.40 |
| one brick thick | m2 | 2.90 | 66.70 | 32.90 | 14.94 | 114.54 |
| one brick thick curved | m2 | 3.50 | 80.50 | 32.90 | 17.01 | 130.41 |
| one and a half brick thick | m2 | 3.60 | 82.80 | 49.35 | 19.82 | 151.97 |
| two brick thick | m2 | 4.30 | 98.90 | 65.80 | 24.71 | 189.41 |
| two brick thick battered | m2 | 4.90 | 112.70 | 65.80 | 26.78 | 205.28 |

walls, facework both sides

| | Unit | Labour hours | Labour cost £ | Materials £ | O & P £ | Total £ |
|---|---|---|---|---|---|---|
| half brick thick | m2 | 1.90 | 43.70 | 16.45 | 9.02 | 69.17 |
| half brick thick overhand | m2 | 2.40 | 55.20 | 16.45 | 10.75 | 82.40 |
| half brick thick against other work | m2 | 2.10 | 48.30 | 16.45 | 9.71 | 74.46 |
| half brick thick curved | m2 | 2.50 | 57.50 | 16.45 | 11.09 | 85.04 |

| | Unit | Labour hours | Labour cost £ | Materials £ | O & P £ | Total £ |
|---|---|---|---|---|---|---|
| one brick thick | m2 | 3.00 | 69.00 | 32.90 | 15.29 | 117.19 |
| one brick thick curved | m2 | 3.60 | 82.80 | 32.90 | 17.36 | 133.06 |
| one and a half brick thick | m2 | 3.70 | 85.10 | 49.35 | 20.17 | 154.62 |
| two brick thick | m2 | 4.40 | 101.20 | 65.80 | 25.05 | 192.05 |
| two brick thick battered | m2 | 5.00 | 115.00 | 65.80 | 27.12 | 207.92 |

Class B engineering bricks basic
price £250 per thousand in
cement mortar (1:3)

walls

| | Unit | Labour hours | Labour cost £ | Materials £ | O & P £ | Total £ |
|---|---|---|---|---|---|---|
| half brick thick | m2 | 1.80 | 41.40 | 20.45 | 9.28 | 71.13 |
| half brick thick overhand | m2 | 2.30 | 52.90 | 20.45 | 11.00 | 84.35 |
| half brick thick against other work | m2 | 2.00 | 46.00 | 20.45 | 9.97 | 76.42 |
| half brick thick curved | m2 | 2.40 | 55.20 | 20.45 | 11.35 | 87.00 |
| one brick thick | m2 | 2.90 | 66.70 | 49.90 | 17.49 | 134.09 |
| one brick thick curved | m2 | 3.50 | 80.50 | 49.90 | 19.56 | 149.96 |
| one and a half brick thick | m2 | 3.60 | 82.80 | 61.35 | 21.62 | 165.77 |
| two brick thick | m2 | 4.30 | 98.90 | 81.90 | 27.12 | 207.92 |
| two brick thick battered | m2 | 4.90 | 112.70 | 81.90 | 29.19 | 223.79 |

walls, facework one side

| | Unit | Labour hours | Labour cost £ | Materials £ | O & P £ | Total £ |
|---|---|---|---|---|---|---|
| half brick thick | m2 | 1.90 | 43.70 | 20.45 | 9.62 | 73.77 |
| half brick thick overhand | m2 | 2.40 | 55.20 | 20.45 | 11.35 | 87.00 |
| half brick thick against other work | m2 | 2.10 | 48.30 | 20.45 | 10.31 | 79.06 |
| half brick thick curved | m2 | 2.50 | 57.50 | 20.45 | 11.69 | 89.64 |
| one brick thick | m2 | 3.00 | 69.00 | 49.90 | 17.84 | 136.74 |
| one brick thick curved | m2 | 3.60 | 82.80 | 49.90 | 19.91 | 152.61 |
| one and a half brick thick | m2 | 3.70 | 85.10 | 61.35 | 21.97 | 168.42 |
| two brick thick | m2 | 4.40 | 101.20 | 81.90 | 27.47 | 210.57 |
| two brick thick battered | m2 | 5.00 | 115.00 | 81.90 | 29.54 | 226.44 |

walls, facework both sides

| | Unit | Labour hours | Labour cost £ | Materials £ | O & P £ | Total £ |
|---|---|---|---|---|---|---|
| half brick thick | m2 | 2.00 | 46.00 | 20.45 | 9.97 | 76.42 |
| half brick thick overhand | m2 | 2.50 | 57.50 | 20.45 | 11.69 | 89.64 |
| half brick thick against other work | m2 | 2.20 | 50.60 | 20.45 | 10.66 | 81.71 |
| half brick thick curved | m2 | 2.60 | 59.80 | 20.45 | 12.04 | 92.29 |

| | Unit | Labour hours | Labour cost £ | Materials £ | O & P £ | Total £ |
|---|---|---|---|---|---|---|
| one brick thick | m2 | 3.10 | 71.30 | 49.90 | 18.18 | 139.38 |
| one brick thick curved | m2 | 3.70 | 85.10 | 49.90 | 20.25 | 155.25 |
| one and a half brick thick | m2 | 3.80 | 87.40 | 61.35 | 22.31 | 171.06 |
| two brick thick | m2 | 4.50 | 103.50 | 81.90 | 27.81 | 213.21 |
| two brick thick battered | m2 | 5.60 | 128.80 | 81.90 | 31.61 | 242.31 |

Class B engineering bricks basic
price £350 per thousand in
cement mortar (1:3)

walls

| | Unit | Labour hours | Labour cost £ | Materials £ | O & P £ | Total £ |
|---|---|---|---|---|---|---|
| half brick thick | m2 | 1.80 | 41.40 | 28.63 | 10.50 | 80.53 |
| half brick thick overhand | m2 | 2.30 | 52.90 | 28.63 | 12.23 | 93.76 |
| half brick thick against other work | m2 | 2.00 | 46.00 | 28.63 | 11.19 | 85.82 |
| half brick thick curved | m2 | 2.40 | 55.20 | 28.63 | 12.57 | 96.40 |
| one brick thick | m2 | 2.90 | 66.70 | 57.26 | 18.59 | 142.55 |
| one brick thick curved | m2 | 3.50 | 80.50 | 57.26 | 20.66 | 158.42 |
| one and a half brick thick | m2 | 3.60 | 82.80 | 85.89 | 25.30 | 193.99 |
| two brick thick | m2 | 4.30 | 98.90 | 114.57 | 32.02 | 245.49 |
| two brick thick battered | m2 | 4.90 | 112.70 | 114.57 | 34.09 | 261.36 |

walls, facework one side

| | Unit | Labour hours | Labour cost £ | Materials £ | O & P £ | Total £ |
|---|---|---|---|---|---|---|
| half brick thick | m2 | 1.90 | 43.70 | 28.63 | 10.85 | 83.18 |
| half brick thick overhand | m2 | 2.40 | 55.20 | 28.63 | 12.57 | 96.40 |
| half brick thick against other work | m2 | 2.10 | 48.30 | 28.63 | 11.54 | 88.47 |
| half brick thick curved | m2 | 2.50 | 57.50 | 28.63 | 12.92 | 99.05 |
| one brick thick | m2 | 3.00 | 69.00 | 57.26 | 18.94 | 145.20 |
| one brick thick curved | m2 | 3.60 | 82.80 | 57.26 | 21.01 | 161.07 |
| one and a half brick thick | m2 | 3.70 | 85.10 | 85.89 | 25.65 | 196.64 |
| two brick thick | m2 | 4.40 | 101.20 | 114.57 | 32.37 | 248.14 |
| two brick thick battered | m2 | 5.00 | 115.00 | 114.57 | 34.44 | 264.01 |

walls, facework both sides

| | Unit | Labour hours | Labour cost £ | Materials £ | O & P £ | Total £ |
|---|---|---|---|---|---|---|
| half brick thick | m2 | 2.00 | 46.00 | 28.63 | 11.19 | 85.82 |
| half brick thick overhand | m2 | 2.50 | 57.50 | 28.63 | 12.92 | 99.05 |
| half brick thick against other work | m2 | 2.20 | 50.60 | 28.63 | 11.88 | 91.11 |
| half brick thick curved | m2 | 2.60 | 59.80 | 28.63 | 13.26 | 101.69 |

|  | Unit | Labour hours | Labour cost £ | Materials £ | O & P £ | Total £ |
|---|---|---|---|---|---|---|
| one brick thick | m2 | 3.10 | 71.30 | 57.26 | 19.28 | 147.84 |
| one brick thick curved | m2 | 3.70 | 85.10 | 57.26 | 21.35 | 163.71 |
| one and a half brick thick | m2 | 3.80 | 87.40 | 85.89 | 25.99 | 199.28 |
| two brick thick | m2 | 4.50 | 103.50 | 114.57 | 32.71 | 250.78 |
| two brick thick battered | m2 | 5.60 | 128.80 | 114.57 | 36.51 | 279.88 |

Facing bricks basic price £250
per 1000 in gauged mortar (1:1:6)

| walls |  |  |  |  |  |  |
|---|---|---|---|---|---|---|
| half brick thick | m2 | 1.70 | 39.10 | 20.30 | 8.91 | 68.31 |
| half brick thick overhand | m2 | 2.20 | 50.60 | 20.30 | 10.64 | 81.54 |
| half brick thick against | | | | | | |
| other work | m2 | 1.90 | 43.70 | 20.30 | 9.60 | 73.60 |
| half brick thick curved | m2 | 2.30 | 52.90 | 20.30 | 10.98 | 84.18 |
| one brick thick | m2 | 2.80 | 64.40 | 40.60 | 15.75 | 120.75 |
| one brick thick curved | m2 | 3.40 | 78.20 | 40.60 | 17.82 | 136.62 |

Extra over for fair face and flush
pointing

| stretcher bond | m2 | 0.60 | 13.80 | 0.10 | 2.09 | 15.99 |
|---|---|---|---|---|---|---|
| flemish bond | m2 | 0.70 | 16.10 | 0.10 | 2.43 | 18.63 |
| margins | m | 0.10 | 2.30 | 0.10 | 0.36 | 2.76 |
| soffits of flat arches | m | 0.65 | 14.95 | 0.10 | 2.26 | 17.31 |
| soffits of curved arches | m | 1.20 | 27.60 | 0.10 | 4.16 | 31.86 |

Isolated casings

| half brick thick, facework all | | | | | | |
|---|---|---|---|---|---|---|
| round, stretcher bond | m2 | 2.30 | 52.90 | 20.40 | 11.00 | 84.30 |
| half brick thick, facework all | | | | | | |
| round, flemish bond | m2 | 2.40 | 55.20 | 20.40 | 11.34 | 86.94 |

Isolated piers

| one brick thick, facework all | | | | | | |
|---|---|---|---|---|---|---|
| round, stretcher bond | m2 | 3.40 | 78.20 | 4.08 | 12.34 | 94.62 |
| one brick thick, facework all | | | | | | |
| round, flemish bond | m2 | 3.50 | 80.50 | 40.80 | 18.20 | 139.50 |

| | Unit | Labour hours | Labour cost £ | Materials £ | O & P £ | Total £ |
|---|---|---|---|---|---|---|
| **Arches** | | | | | | |
| 215mm high 102mm wide | m | 1.38 | 31.74 | 4.36 | 5.42 | 41.52 |
| 215mm high 215mm wide | m | 2.05 | 47.15 | 7.34 | 8.17 | 62.66 |
| 215mm high, 102mm wide, segmental | m | 2.76 | 63.48 | 8.27 | 10.76 | 82.51 |
| 215mm high, 215mm wide, segmental | m | 3.72 | 85.56 | 11.56 | 14.57 | 111.69 |
| **Projections** | | | | | | |
| 112mm high, 225mm wide, stretcher bond | m | 0.48 | 11.04 | 5.14 | 2.43 | 18.61 |
| 112mm high, 225mm wide, flemish bond | m | 0.60 | 13.80 | 5.14 | 2.84 | 21.78 |
| 225mm high, 225mm wide, flemish bond | m | 0.85 | 19.55 | 10.28 | 4.47 | 34.30 |
| 112mm high, 328mm wide, stretcher bond | m | 0.75 | 17.25 | 7.17 | 3.66 | 28.08 |
| 112mm high, 328mm wide, flemish bond | m | 0.95 | 21.85 | 7.17 | 4.35 | 33.37 |
| **Facing bricks basic price £400 per 1000 in gauged mortar (1:1:6)** | | | | | | |
| walls | | | | | | |
| half brick thick | m2 | 1.70 | 39.10 | 28.55 | 10.15 | 77.80 |
| half brick thick overhand | m2 | 2.20 | 50.60 | 28.55 | 11.87 | 91.02 |
| half brick thick against other work | m2 | 1.90 | 43.70 | 28.55 | 10.84 | 83.09 |
| half brick thick curved | m2 | 2.30 | 52.90 | 28.55 | 12.22 | 93.67 |
| one brick thick | m2 | 2.80 | 64.40 | 57.10 | 18.23 | 139.73 |
| one brick thick curved | m2 | 3.40 | 78.20 | 57.10 | 20.30 | 155.60 |
| **Extra over for fair face and flush pointing** | | | | | | |
| stretcher bond | m2 | 0.60 | 13.80 | 0.10 | 2.09 | 15.99 |
| flemish bond | m2 | 0.70 | 16.10 | 0.10 | 2.43 | 18.63 |
| margins | m | 0.10 | 2.30 | 0.10 | 0.36 | 2.76 |

|  | Unit | Labour hours | Labour cost £ | Materials £ | O & P £ | Total £ |
|---|---|---|---|---|---|---|
| soffits of flat arches | m | 0.65 | 14.95 | 0.10 | 2.26 | 17.31 |
| soffits of curved arches | m | 1.20 | 27.60 | 0.10 | 4.16 | 31.86 |

Isolated casings

|  | Unit | Labour hours | Labour cost £ | Materials £ | O & P £ | Total £ |
|---|---|---|---|---|---|---|
| half brick thick, facework all round, stretcher bond | m2 | 2.30 | 52.90 | 28.65 | 12.23 | 93.78 |
| half brick thick, facework all round, flemish bond | m2 | 2.40 | 55.20 | 28.65 | 12.58 | 96.43 |

Isolated piers

|  | Unit | Labour hours | Labour cost £ | Materials £ | O & P £ | Total £ |
|---|---|---|---|---|---|---|
| one brick thick, facework all round, stretcher bond | m2 | 3.40 | 78.20 | 57.30 | 20.33 | 155.83 |
| one brick thick, facework all round, flemish bond | m2 | 3.50 | 80.50 | 57.30 | 20.67 | 158.47 |

Arches

|  | Unit | Labour hours | Labour cost £ | Materials £ | O & P £ | Total £ |
|---|---|---|---|---|---|---|
| 215mm high 102mm wide | m | 1.38 | 31.74 | 6.98 | 5.81 | 44.53 |
| 215mm high 215mm wide | m | 2.05 | 47.15 | 11.74 | 8.83 | 67.72 |
| 215mm high, 102mm wide, segmental | m | 2.76 | 63.48 | 13.23 | 11.51 | 88.22 |
| 215mm high, 215mm wide, segmental | m | 3.72 | 85.56 | 18.50 | 15.61 | 119.67 |

Projections

|  | Unit | Labour hours | Labour cost £ | Materials £ | O & P £ | Total £ |
|---|---|---|---|---|---|---|
| 112mm high, 225mm wide, stretcher bond | m | 0.48 | 11.04 | 8.22 | 2.89 | 22.15 |
| 112mm high, 225mm wide, flemish bond | m | 0.60 | 13.80 | 8.22 | 3.30 | 25.32 |
| 225mm high, 225mm wide, flemish bond | m | 0.85 | 19.55 | 16.38 | 5.39 | 41.32 |
| 112mm high, 328mm wide, stretcher bond | m | 0.75 | 17.25 | 11.47 | 4.31 | 33.03 |
| 112mm high, 328mm wide, flemish bond | m | 0.95 | 21.85 | 11.47 | 5.00 | 38.32 |

| | Unit | Labour hours | Labour cost £ | Materials £ | O & P £ | Total £ |
|---|---|---|---|---|---|---|
| **Facing bricks basic price £500 per 1000 in gauged mortar (1:1:6)** | | | | | | |
| walls | | | | | | |
| half brick thick | m2 | 1.70 | 39.10 | 34.85 | 11.09 | 85.04 |
| half brick thick overhand | m2 | 2.20 | 50.60 | 34.85 | 12.82 | 98.27 |
| half brick thick against other work | m2 | 1.90 | 43.70 | 34.85 | 11.78 | 90.33 |
| half brick thick curved | m2 | 2.30 | 52.90 | 34.85 | 13.16 | 100.91 |
| one brick thick | m2 | 2.80 | 64.40 | 69.70 | 20.12 | 154.22 |
| one brick thick curved | m2 | 3.40 | 78.20 | 69.70 | 22.19 | 170.09 |
| **Extra over for fair face and flush pointing** | | | | | | |
| stretcher bond | m2 | 0.60 | 13.80 | 0.10 | 2.09 | 15.99 |
| flemish bond | m2 | 0.70 | 16.10 | 0.10 | 2.43 | 18.63 |
| margins | m | 0.10 | 2.30 | 0.10 | 0.36 | 2.76 |
| soffits of flat arches | m | 0.65 | 14.95 | 0.10 | 2.26 | 17.31 |
| soffits of curved arches | m | 1.20 | 27.60 | 0.10 | 4.16 | 31.86 |
| **Isolated casings** | | | | | | |
| half brick thick, facework all round, stretcher bond | m2 | 2.30 | 52.90 | 34.85 | 13.16 | 100.91 |
| half brick thick, facework all round, flemish bond | m2 | 2.40 | 55.20 | 34.85 | 13.51 | 103.56 |
| **Isolated piers** | | | | | | |
| one brick thick, facework all round, stretcher bond | m2 | 3.40 | 78.20 | 69.70 | 22.19 | 170.09 |
| one brick thick, facework all round, flemish bond | m2 | 3.50 | 80.50 | 69.70 | 22.53 | 172.73 |
| **Arches** | | | | | | |
| 215mm high 102mm wide | m | 1.38 | 31.74 | 8.73 | 6.07 | 46.54 |
| 215mm high 215mm wide | m | 2.05 | 47.15 | 14.68 | 9.27 | 71.10 |

| | Unit | Labour hours | Labour cost £ | Materials £ | O & P £ | Total £ |
|---|---|---|---|---|---|---|
| 215mm high, 102mm wide, segmental | m | 2.76 | 63.48 | 16.54 | 12.00 | 92.02 |
| 215mm high, 215mm wide, segmental | m | 3.72 | 85.56 | 23.13 | 16.30 | 124.99 |

Projections

| | Unit | Labour hours | Labour cost £ | Materials £ | O & P £ | Total £ |
|---|---|---|---|---|---|---|
| 112mm high, 225mm wide, stretcher bond | m | 0.48 | 11.04 | 10.28 | 3.20 | 24.52 |
| 112mm high, 225mm wide, flemish bond | m | 0.60 | 13.80 | 10.28 | 3.61 | 27.69 |
| 225mm high, 225mm wide, flemish bond | m | 0.85 | 19.55 | 16.38 | 5.39 | 41.32 |
| 112mm high, 328mm wide, stretcher bond | m | 0.75 | 17.25 | 20.48 | 5.66 | 43.39 |
| 112mm high, 328mm wide, flemish bond | m | 0.95 | 21.85 | 14.34 | 5.43 | 41.62 |

## Blockwork

Concrete dense aggregate blocks
in gauged mortar (1:1:6)

| | Unit | Labour hours | Labour cost £ | Materials £ | O & P £ | Total £ |
|---|---|---|---|---|---|---|
| walls and partitions | | | | | | |
| 75mm thick, solid | m2 | 0.98 | 22.54 | 8.25 | 4.62 | 35.41 |
| 100mm thick, solid | m2 | 1.14 | 26.22 | 8.89 | 5.27 | 40.38 |
| 140mm thick, solid | m2 | 1.32 | 30.36 | 16.68 | 7.06 | 54.10 |
| 140mm thick, hollow | m2 | 1.43 | 32.89 | 16.39 | 7.39 | 56.67 |
| 190mm thick, hollow | m2 | 1.76 | 40.48 | 18.63 | 8.87 | 67.98 |
| 215mm thick, hollow | m2 | 2.02 | 46.46 | 18.84 | 9.80 | 75.10 |
| skins of hollow walls | | | | | | |
| 75mm thick, solid | m2 | 1.12 | 25.76 | 8.25 | 5.10 | 39.11 |
| 100mm thick, solid | m2 | 1.24 | 28.52 | 8.89 | 5.61 | 43.02 |
| 140mm thick, solid | m2 | 1.27 | 29.21 | 16.68 | 6.88 | 52.77 |
| 215mm thick, hollow | m2 | 2.15 | 49.45 | 18.84 | 10.24 | 78.53 |

| | Unit | Labour hours | Labour cost £ | Materials £ | O & P £ | Total £ |
|---|---|---|---|---|---|---|
| **piers and chimney stacks** | | | | | | |
| 75mm thick, solid | m2 | 1.28 | 29.44 | 8.25 | 5.65 | 43.34 |
| 100mm thick, solid | m2 | 1.40 | 32.20 | 8.89 | 6.16 | 47.25 |
| 140mm thick, solid | m2 | 1.68 | 38.64 | 16.68 | 8.30 | 63.62 |
| **isolated casings** | | | | | | |
| 75mm thick, solid | m2 | 1.42 | 32.66 | 8.25 | 6.14 | 47.05 |
| 100mm thick, solid | m2 | 1.90 | 43.70 | 8.89 | 7.89 | 60.48 |
| 140mm thick, solid | m2 | 2.15 | 49.45 | 16.68 | 9.92 | 76.05 |
| **Extra over for fair face and flush pointing** | | | | | | |
| one side | m2 | 0.16 | 3.68 | 0.00 | 0.55 | 4.23 |
| two sides | m2 | 0.33 | 7.59 | 0.00 | 1.14 | 8.73 |
| **Bonding ends of blockwork to brickwork** | | | | | | |
| 75mm thick, solid | m2 | 0.30 | 6.90 | 0.00 | 1.04 | 7.94 |
| 100mm thick, solid | m2 | 0.38 | 8.74 | 0.00 | 1.31 | 10.05 |
| 140mm thick, solid | m2 | 0.44 | 10.12 | 0.00 | 1.52 | 11.64 |
| 140mm thick, hollow | m2 | 0.48 | 11.04 | 0.00 | 1.66 | 12.70 |
| 190mm thick, hollow | m2 | 0.55 | 12.65 | 0.00 | 1.90 | 14.55 |
| 215mm thick, hollow | m2 | 0.62 | 14.26 | 0.00 | 2.14 | 16.40 |
| **Close cavities at ends of walls and jambs with 100mm thick blockwork and bitumen damp proof course bedded in mortar (1:1:6)** | m | 0.20 | 4.60 | 1.71 | 0.95 | 7.26 |
| **Close cavities at tops of walls with blocks laid flat in mortar (1:1:6)** | | | | | | |
| 100mm thick, solid | m | 0.24 | 5.52 | 3.28 | 1.32 | 10.12 |
| 140mm thick, solid | m | 0.28 | 6.44 | 3.70 | 1.52 | 11.66 |

| | Unit | Labour hours | Labour cost £ | Materials £ | O & P £ | Total £ |
|---|---|---|---|---|---|---|
| **Lightweight aerated concrete blocks in mortar (1:1:6)** | | | | | | |
| | | | | | | |
| walls and partitions | | | | | | |
| 75mm thick | m2 | 0.90 | 20.70 | 8.42 | 4.37 | 33.49 |
| 100mm thick | m2 | 1.05 | 24.15 | 9.79 | 5.09 | 39.03 |
| 125mm thick | m2 | 1.20 | 27.60 | 11.26 | 5.83 | 44.69 |
| 140mm thick | m2 | 1.26 | 28.98 | 16.78 | 6.86 | 52.62 |
| 150mm thick | m2 | 1.34 | 30.82 | 18.80 | 7.44 | 57.06 |
| 190mm thick | m2 | 1.60 | 36.80 | 20.52 | 8.60 | 65.92 |
| 215mm thick | m2 | 2.00 | 46.00 | 23.65 | 10.45 | 80.10 |
| | | | | | | |
| skins of hollow walls | | | | | | |
| 75mm thick | m2 | 1.00 | 23.00 | 8.42 | 4.71 | 36.13 |
| 100mm thick | m2 | 1.15 | 26.45 | 9.79 | 5.44 | 41.68 |
| 125mm thick | m2 | 1.30 | 29.90 | 11.26 | 6.17 | 47.33 |
| 150mm thick | m2 | 1.44 | 33.12 | 18.80 | 7.79 | 59.71 |
| piers and chimney stacks | | | | | | |
| 140mm thick | m2 | 1.44 | 33.12 | 16.78 | 7.49 | 57.39 |
| 190mm thick | m2 | 1.52 | 34.96 | 20.52 | 8.32 | 63.80 |
| 215mm thick | m2 | 1.62 | 37.26 | 23.65 | 9.14 | 70.05 |
| isolated casings | | | | | | |
| 75mm thick | m2 | 1.60 | 36.80 | 8.42 | 6.78 | 52.00 |
| 100mm thick | m2 | 1.92 | 44.16 | 9.79 | 8.09 | 62.04 |
| 140mm thick | m2 | 2.07 | 47.61 | 16.78 | 9.66 | 74.05 |
| **Extra over for fair face and flush pointing** | | | | | | |
| | | | | | | |
| one side | m2 | 0.16 | 3.68 | 0.00 | 0.55 | 4.23 |
| two sides | m2 | 0.33 | 7.59 | 0.00 | 1.14 | 8.73 |

| | Unit | Labour hours | Labour cost £ | Materials £ | O & P £ | Total £ |
|---|---|---|---|---|---|---|
| **Bonding ends of blockwork to brickwork** | | | | | | |
| 75mm thick | m | 0.25 | 5.75 | 0.00 | 0.86 | 6.61 |
| 100mm thick | m | 0.32 | 7.36 | 0.00 | 1.10 | 8.46 |
| 125mm thick | m | 0.36 | 8.28 | 0.00 | 1.24 | 9.52 |
| 140mm thick | m | 0.40 | 9.20 | 0.00 | 1.38 | 10.58 |
| 150mm thick | m | 0.44 | 10.12 | 0.00 | 1.52 | 11.64 |
| 190mm thick | m | 0.50 | 11.50 | 0.00 | 1.73 | 13.23 |
| 215mm thick | m | 0.55 | 12.65 | 0.00 | 1.90 | 14.55 |
| 255mm thick | m | 0.58 | 13.34 | 0.00 | 2.00 | 15.34 |
| **Close cavities at ends of walls and jambs with 100mm thick blockwork and bitumen damp proof course** | | | | | | |
| bedded in mortar (1:1:6) | m | 0.20 | 4.60 | 1.86 | 0.97 | 7.43 |
| **Close cavities at tops of walls with blocks laid flat in mortar (1:1:6)** | | | | | | |
| 100mm thick, solid | m | 0.95 | 21.85 | 1.63 | 3.52 | 27.00 |
| 140mm thick, solid | m | 1.05 | 24.15 | 2.79 | 4.04 | 30.98 |
| **Lightweight aerated concrete blocks, smooth faced, in mortar (1:1:6)** | | | | | | |
| walls and partitions | | | | | | |
| 100mm thick | m2 | 1.15 | 26.45 | 19.14 | 6.84 | 52.43 |
| 140mm thick | m2 | 1.36 | 31.28 | 26.86 | 8.72 | 66.86 |
| 190mm thick | m2 | 1.70 | 39.10 | 35.66 | 11.21 | 85.97 |
| 215mm thick | m2 | 2.10 | 48.30 | 45.59 | 14.08 | 107.97 |
| piers and chimney stacks | | | | | | |
| 140mm thick | m2 | 1.54 | 35.42 | 26.86 | 9.34 | 71.62 |
| 190mm thick | m2 | 1.62 | 37.26 | 35.66 | 10.94 | 83.86 |
| 215mm thick | m2 | 1.72 | 39.56 | 45.59 | 12.77 | 97.92 |

| | Unit | Labour hours | Labour cost £ | Materials £ | O & P £ | Total £ |
|---|---|---|---|---|---|---|
| isolated casings | | | | | | |
| 100mm thick | m2 | 2.02 | 46.46 | 19.14 | 9.84 | 75.44 |
| 140mm thick | m2 | 2.16 | 49.68 | 26.86 | 11.48 | 88.02 |
| **Extra over for fair face and flush pointing** | | | | | | |
| one side | m2 | 0.16 | 3.68 | 0.00 | 0.55 | 4.23 |
| two sides | m2 | 0.33 | 7.59 | 0.00 | 1.14 | 8.73 |

## Damp proof courses

Hessian based bitumen damp proof course in gauged mortar (1:1:6)

| | Unit | Labour hours | Labour cost £ | Materials £ | O & P £ | Total £ |
|---|---|---|---|---|---|---|
| horizontal, width | | | | | | |
| over 225mm | m2 | 0.35 | 8.05 | 10.02 | 2.71 | 20.78 |
| 112mm | m | 0.05 | 1.15 | 1.43 | 0.39 | 2.97 |
| vertical, width | | | | | | |
| over 225mm | m2 | 0.40 | 9.20 | 10.02 | 2.88 | 22.10 |
| 112mm | m | 0.07 | 1.61 | 1.43 | 0.46 | 3.50 |

Fibre based damp proof course in gauged mortar (1:1:6)

| | Unit | Labour hours | Labour cost £ | Materials £ | O & P £ | Total £ |
|---|---|---|---|---|---|---|
| horizontal, width | | | | | | |
| over 225mm | m2 | 0.40 | 9.20 | 7.00 | 2.43 | 18.63 |
| 112mm | m | 0.05 | 1.15 | 0.95 | 0.32 | 2.42 |
| vertical, width | | | | | | |
| over 225mm | m2 | 0.45 | 10.35 | 7.00 | 2.60 | 19.95 |
| 112mm | m | 0.07 | 1.61 | 0.95 | 0.38 | 2.94 |

| | Unit | Labour hours | Labour cost £ | Materials £ | O & P £ | Total £ |
|---|---|---|---|---|---|---|
| **Pitch polymer damp proof course in gauged mortar (1:1:6)** | | | | | | |
| horizontal, width | | | | | | |
| over 225mm | m2 | 0.40 | 9.20 | 8.38 | 2.64 | 20.22 |
| 112mm | m | 0.05 | 1.15 | 0.99 | 0.32 | 2.46 |
| vertical, width | | | | | | |
| over 225mm | m2 | 0.45 | 10.35 | 8.38 | 2.81 | 21.54 |
| 112mm | m | 0.07 | 1.61 | 0.99 | 0.39 | 2.99 |
| **Bitumen damp proof course with lead core in mortar (1:1:6)** | | | | | | |
| horizontal, width | | | | | | |
| over 225mm | m2 | 0.50 | 11.50 | 22.18 | 5.05 | 38.73 |
| 112mm | m | 0.06 | 1.38 | 2.72 | 0.62 | 4.72 |
| vertical, width | | | | | | |
| over 225mm | m2 | 0.55 | 12.65 | 22.18 | 5.22 | 40.05 |
| 112mm | m | 0.08 | 1.84 | 2.72 | 0.68 | 5.24 |
| **Two courses of slates bedded in cement mortar (1:3)** | | | | | | |
| horizontal, width | | | | | | |
| over 225mm | m2 | 0.87 | 20.01 | 35.95 | 8.39 | 64.35 |
| 112mm | m | 0.33 | 7.59 | 3.47 | 1.66 | 12.72 |
| vertical, width | | | | | | |
| over 225mm | m2 | 0.95 | 21.85 | 35.95 | 8.67 | 66.47 |
| 112mm | m | 0.35 | 8.05 | 3.47 | 1.73 | 13.25 |
| **Three courses of waterproofing liquid brush on concrete surfaces** | | | | | | |
| vertically | m2 | 0.45 | 10.35 | 9.16 | 2.93 | 22.44 |
| horizontally | m2 | 0.50 | 11.50 | 9.16 | 3.10 | 23.76 |

| | Unit | Labour hours | Labour cost £ | Materials £ | O & P £ | Total £ |
|---|---|---|---|---|---|---|
| **Sundries** | | | | | | |
| Expanded polystyrene cavity wall insulation sheeting, thickness | | | | | | |
| 25mm | m2 | 0.20 | 4.60 | 5.51 | 1.52 | 11.63 |
| 50mm | m2 | 0.22 | 5.06 | 8.46 | 2.03 | 15.55 |
| Galvanised steel brick reinforcement, width | | | | | | |
| 65mm | m | 0.07 | 1.61 | 0.43 | 0.31 | 2.35 |
| 115mm | m | 0.09 | 2.07 | 0.52 | 0.39 | 2.98 |
| 175mm | m | 0.12 | 2.76 | 0.84 | 0.54 | 4.14 |
| 225mm | m | 0.15 | 3.45 | 1.20 | 0.70 | 5.35 |
| Point frames with mastic | | | | | | |
| one side | m | 0.14 | 3.22 | 0.63 | 0.58 | 4.43 |
| both sides | m | 0.22 | 5.06 | 1.25 | 0.95 | 7.26 |
| Point frames with polysulphide sealant | | | | | | |
| one side | m | 0.14 | 3.22 | 2.05 | 0.79 | 6.06 |
| both sides | m | 0.22 | 5.06 | 4.11 | 1.38 | 10.55 |
| Rake out joints for flashings and point up on completion | | | | | | |
| horizontal | m | 0.09 | 2.07 | 0.24 | 0.35 | 2.66 |
| stepped | m | 0.18 | 4.14 | 0.29 | 0.66 | 5.09 |

| | Unit | Labour hours | Labour cost £ | Materials £ | O & P £ | Total £ |
|---|---|---|---|---|---|---|
| **MASONRY** | | | | | | |
| **Walling** | | | | | | |
| Random rubble walling, laid dry, thickness | | | | | | |
| 300mm | m2 | 3.00 | 48.00 | 55.70 | 15.56 | 119.26 |
| 450mm | m2 | 3.50 | 56.00 | 83.47 | 20.92 | 160.39 |
| 500mm | m2 | 3.75 | 60.00 | 94.16 | 23.12 | 177.28 |
| Random rubble walling, laid dry, battered one side, thickness | | | | | | |
| 300mm | m2 | 3.25 | 52.00 | 55.70 | 16.16 | 123.86 |
| 450mm | m2 | 3.75 | 60.00 | 83.47 | 21.52 | 164.99 |
| 500mm | m2 | 4.00 | 64.00 | 94.16 | 23.72 | 181.88 |
| Random rubble walling, laid dry, battered both sides, thickness | | | | | | |
| 300mm | m2 | 3.50 | 56.00 | 55.70 | 16.76 | 128.46 |
| 450mm | m2 | 4.00 | 64.00 | 83.47 | 22.12 | 169.59 |
| 500mm | m2 | 4.25 | 68.00 | 94.16 | 24.32 | 186.48 |
| Random rubble walling, laid in gauged mortar (1:1:6), thickness | | | | | | |
| 300mm | m2 | 3.20 | 51.20 | 59.85 | 16.66 | 127.71 |
| 450mm | m2 | 3.70 | 59.20 | 89.78 | 22.35 | 171.33 |
| 500mm | m2 | 3.90 | 62.40 | 100.80 | 24.48 | 187.68 |
| Random rubble walling, laid in gauged mortar (1:1:6), battered one side, thickness | | | | | | |
| 300mm | m2 | 3.45 | 55.20 | 59.85 | 17.26 | 132.31 |
| 450mm | m2 | 4.00 | 64.00 | 89.78 | 23.07 | 176.85 |
| 500mm | m2 | 4.15 | 66.40 | 100.80 | 25.08 | 192.28 |

| | Unit | Labour hours | Labour cost £ | Materials £ | O & P £ | Total £ |
|---|---|---|---|---|---|---|
| **Random rubble walling, laid in gauged mortar (1:1:6), battered both sides, thickness** | | | | | | |
| 300mm | m2 | 3.95 | 63.20 | 59.85 | 18.46 | 141.51 |
| 450mm | m2 | 4.50 | 72.00 | 89.78 | 24.27 | 186.05 |
| 500mm | m2 | 4.65 | 74.40 | 100.80 | 26.28 | 201.48 |
| **Irregular coursed rubble walling, laid in gauged mortar (1:1:6), thickness** | | | | | | |
| 300mm | m2 | 2.10 | 33.60 | 59.85 | 14.02 | 107.47 |
| 450mm | m2 | 2.90 | 46.40 | 89.78 | 20.43 | 156.61 |
| 500mm | m2 | 3.40 | 54.40 | 100.80 | 23.28 | 178.48 |
| **Coursed rubble walling, laid in gauged mortar (1:1:6), thickness** | | | | | | |
| 300mm | m2 | 2.40 | 38.40 | 59.85 | 14.74 | 112.99 |
| 450mm | m2 | 3.20 | 51.20 | 89.78 | 21.15 | 162.13 |
| 500mm | m2 | 3.70 | 59.20 | 100.80 | 24.00 | 184.00 |
| **Fair raking cutting on stone walling, thickness** | | | | | | |
| 300mm | m | 0.80 | 12.80 | 0.00 | 1.92 | 14.72 |
| 450mm | m | 1.20 | 19.20 | 0.00 | 2.88 | 22.08 |
| 500mm | m | 1.80 | 28.80 | 0.00 | 4.32 | 33.12 |
| **Form level bed on stone walling, thickness** | | | | | | |
| 300mm | m | 0.20 | 3.20 | 0.00 | 0.48 | 3.68 |
| 450mm | m | 0.35 | 5.60 | 0.00 | 0.84 | 6.44 |
| 500mm | m | 0.50 | 8.00 | 0.00 | 1.20 | 9.20 |

| | Unit | Labour hours | Labour cost £ | Materials £ | O & P £ | Total £ |
|---|---|---|---|---|---|---|
| **Sundries** | | | | | | |
| Form holes for pipes up to 50mm diameter through stone walling, thickness | | | | | | |
| 300mm | m | 1.20 | 19.20 | 0.00 | 2.88 | 22.08 |
| 450mm | m | 1.80 | 28.80 | 0.00 | 4.32 | 33.12 |
| 500mm | m | 2.65 | 42.40 | 0.00 | 6.36 | 48.76 |
| Form holes for pipes 50-100mm diameter through stone walling, thickness | | | | | | |
| 300mm | m | 1.40 | 22.40 | 0.00 | 3.36 | 25.76 |
| 450mm | m | 2.00 | 32.00 | 0.00 | 4.80 | 36.80 |
| 500mm | m | 2.85 | 45.60 | 0.00 | 6.84 | 52.44 |
| Form holes for pipes over 100mm diameter through stone walling, thickness | | | | | | |
| 300mm | m | 2.40 | 38.40 | 0.00 | 5.76 | 44.16 |
| 450mm | m | 3.00 | 48.00 | 0.00 | 7.20 | 55.20 |
| 500mm | m | 3.85 | 61.60 | 0.00 | 9.24 | 70.84 |
| Build in ends of steel sections in stone walling, size | | | | | | |
| not exceeding 250mm deep | m | 0.70 | 11.20 | 0.00 | 1.68 | 12.88 |
| 250-500mm deep | m | 0.90 | 14.40 | 0.00 | 2.16 | 16.56 |
| over 500mm deep | m | 1.10 | 17.60 | 0.00 | 2.64 | 20.24 |
| Mortices in stone walling, size | | | | | | |
| 50 × 50 × 100mm | nr | 0.60 | 9.60 | 0.00 | 1.44 | 11.04 |
| 50 × 50 × 150mm | nr | 0.65 | 10.40 | 0.00 | 1.56 | 11.96 |
| 75 × 75 × 100mm | nr | 0.70 | 11.20 | 0.00 | 1.68 | 12.88 |
| 75 × 75 × 150mm | nr | 0.75 | 12.00 | 0.00 | 1.80 | 13.80 |

| | Unit | Labour hours | Labour cost £ | Materials £ | O & P £ | Total £ |
|---|---|---|---|---|---|---|
| **Grout up mortices in cement mortar (1:3)** | | | | | | |
| 50 × 50 × 100mm | nr | 0.10 | 1.60 | 0.15 | 0.26 | 2.01 |
| 50 × 50 × 150mm | nr | 0.15 | 2.40 | 0.22 | 0.39 | 3.01 |
| 75 × 75 × 100mm | nr | 0.15 | 2.40 | 0.34 | 0.41 | 3.15 |
| 75 × 75 × 150mm | nr | 0.20 | 3.20 | 0.40 | 0.54 | 4.14 |

| | Unit | Labour hours | Labour cost £ | Materials £ | O & P £ | Total £ |
|---|---|---|---|---|---|---|

## CARPENTRY AND JOINERY

### Sawn softwood, untreated

Floors

| | Unit | Labour hours | Labour cost £ | Materials £ | O & P £ | Total £ |
|---|---|---|---|---|---|---|
| 50 × 100mm | m | 0.16 | 2.56 | 1.47 | 0.60 | 4.63 |
| 50 × 125mm | m | 0.17 | 2.72 | 1.81 | 0.68 | 5.21 |
| 50 × 150mm | m | 0.18 | 2.88 | 2.23 | 0.77 | 5.88 |
| 75 × 125mm | m | 0.17 | 2.72 | 3.13 | 0.88 | 6.73 |
| 75 × 150mm | m | 0.18 | 2.88 | 3.66 | 0.98 | 7.52 |
| 75 × 200mm | m | 0.19 | 3.04 | 4.27 | 1.10 | 8.41 |
| 75 × 225mm | m | 0.20 | 3.20 | 5.34 | 1.28 | 9.82 |

Partitions

| | Unit | Labour hours | Labour cost £ | Materials £ | O & P £ | Total £ |
|---|---|---|---|---|---|---|
| 38 × 75mm | m | 0.15 | 2.40 | 1.31 | 0.56 | 4.27 |
| 38 × 100mm | m | 0.16 | 2.56 | 1.45 | 0.60 | 4.61 |
| 50 × 75mm | m | 0.16 | 2.56 | 1.43 | 0.60 | 4.59 |
| 50 × 100mm | m | 0.17 | 2.72 | 1.47 | 0.63 | 4.82 |

Flat roofs

| | Unit | Labour hours | Labour cost £ | Materials £ | O & P £ | Total £ |
|---|---|---|---|---|---|---|
| 38 × 100mm | m | 0.10 | 1.60 | 1.45 | 0.46 | 3.51 |
| 50 × 75mm | m | 0.11 | 1.76 | 1.43 | 0.48 | 3.67 |
| 50 × 100mm | m | 0.12 | 1.92 | 1.47 | 0.51 | 3.90 |
| 50 × 125mm | m | 0.13 | 2.08 | 1.81 | 0.58 | 4.47 |
| 50 × 150mm | m | 0.15 | 2.40 | 1.87 | 0.64 | 4.91 |
| 75 × 125mm | m | 0.16 | 2.56 | 3.13 | 0.85 | 6.54 |

Pitched roofs

| | Unit | Labour hours | Labour cost £ | Materials £ | O & P £ | Total £ |
|---|---|---|---|---|---|---|
| 38 × 100mm | m | 0.14 | 2.24 | 1.45 | 0.55 | 4.24 |
| 50 × 75mm | m | 0.15 | 2.40 | 1.43 | 0.57 | 4.40 |
| 50 × 100mm | m | 0.16 | 2.56 | 1.47 | 0.60 | 4.63 |
| 50 × 125mm | m | 0.17 | 2.72 | 1.81 | 0.68 | 5.21 |
| 50 × 150mm | m | 0.19 | 3.04 | 1.87 | 0.74 | 5.65 |
| 75 × 125mm | m | 0.20 | 3.20 | 3.13 | 0.95 | 7.28 |
| 75 × 150mm | m | 0.23 | 3.70 | 3.61 | 1.10 | 8.40 |

| | Unit | Labour hours | Labour cost £ | Materials £ | O & P £ | Total £ |
|---|---|---|---|---|---|---|
| **Kerbs and bearers** | | | | | | |
| 25 × 75mm | m | 0.13 | 2.08 | 0.80 | 0.43 | 3.31 |
| 25 × 100mm | m | 0.14 | 2.24 | 0.89 | 0.47 | 3.60 |
| 25 × 150mm | m | 0.17 | 2.72 | 1.26 | 0.60 | 4.58 |
| 38 × 75mm | m | 0.15 | 2.40 | 1.04 | 0.52 | 3.96 |
| 38 × 100mm | m | 0.16 | 2.56 | 1.45 | 0.60 | 4.61 |
| 38 × 150mm | m | 0.18 | 2.88 | 1.53 | 0.66 | 5.07 |
| 50 × 50mm | m | 0.16 | 2.56 | 1.11 | 0.55 | 4.22 |
| 50 × 75mm | m | 0.17 | 2.72 | 1.43 | 0.62 | 4.77 |
| 50 × 100mm | m | 0.18 | 2.88 | 1.47 | 0.65 | 5.00 |
| 75 × 75mm | m | 0.20 | 3.20 | 1.60 | 0.72 | 5.52 |
| 75 × 100mm | m | 0.22 | 3.52 | 2.14 | 0.85 | 6.51 |
| 75 × 125mm | m | 0.24 | 3.84 | 3.13 | 1.05 | 8.02 |
| **Solid strutting** | | | | | | |
| 38 × 100mm | m | 0.40 | 6.40 | 1.45 | 1.18 | 9.03 |
| 50 × 100mm | m | 0.44 | 7.04 | 1.47 | 1.28 | 9.79 |
| 50 × 125mm | m | 0.48 | 7.68 | 1.66 | 1.40 | 10.74 |
| 50 × 150mm | m | 0.52 | 8.32 | 1.81 | 1.52 | 11.65 |
| **Herringbone strutting, 50 × 50mm to joists, depth** | | | | | | |
| 125mm | m | 0.60 | 9.60 | 1.56 | 1.67 | 12.83 |
| 150mm | m | 0.60 | 9.60 | 1.60 | 1.68 | 12.88 |
| 175mm | m | 0.60 | 9.60 | 1.66 | 1.69 | 12.95 |
| 240mm | m | 0.60 | 9.60 | 1.71 | 1.70 | 13.01 |
| **Sawn softwood, impregnated** | | | | | | |
| Floors | | | | | | |
| 50 × 100mm | m | 0.16 | 2.56 | 1.56 | 0.62 | 4.74 |
| 50 × 125mm | m | 0.17 | 2.72 | 1.90 | 0.69 | 5.31 |
| 50 × 150mm | m | 0.18 | 2.88 | 2.32 | 0.78 | 5.98 |
| 75 × 125mm | m | 0.17 | 2.72 | 3.28 | 0.90 | 6.90 |
| 75 × 150mm | m | 0.18 | 2.88 | 3.80 | 1.00 | 7.68 |

|  | Unit | Labour hours | Labour cost £ | Materials £ | O & P £ | Total £ |
|---|---|---|---|---|---|---|
| 75 × 200mm | m | 0.19 | 3.04 | 4.31 | 1.10 | 8.45 |
| 75 × 225mm | m | 0.20 | 3.20 | 5.47 | 1.30 | 9.97 |

Partitions

| | | | | | | |
|---|---|---|---|---|---|---|
| 38 × 75mm | m | 0.15 | 2.40 | 1.36 | 0.56 | 4.32 |
| 38 × 100mm | m | 0.16 | 2.56 | 1.48 | 0.61 | 4.65 |
| 50 × 75mm | m | 0.16 | 2.56 | 1.49 | 0.61 | 4.66 |
| 50 × 100mm | m | 0.17 | 2.72 | 1.56 | 0.64 | 4.92 |

Flat roofs

| | | | | | | |
|---|---|---|---|---|---|---|
| 38 × 100mm | m | 0.10 | 1.60 | 1.48 | 0.46 | 3.54 |
| 50 × 75mm | m | 0.11 | 1.76 | 1.49 | 0.49 | 3.74 |
| 50 × 100mm | m | 0.12 | 1.92 | 1.56 | 0.52 | 4.00 |
| 50 × 125mm | m | 0.13 | 2.08 | 1.90 | 0.60 | 4.58 |
| 50 × 150mm | m | 0.15 | 2.40 | 2.32 | 0.71 | 5.43 |
| 75 × 125mm | m | 0.16 | 2.56 | 3.29 | 0.88 | 6.73 |

Pitched roofs

| | | | | | | |
|---|---|---|---|---|---|---|
| 38 × 100mm | m | 0.14 | 2.24 | 1.48 | 0.56 | 4.28 |
| 50 × 75mm | m | 0.15 | 2.40 | 1.44 | 0.58 | 4.42 |
| 50 × 100mm | m | 0.16 | 2.56 | 1.56 | 0.62 | 4.74 |
| 50 × 125mm | m | 0.17 | 2.72 | 1.90 | 0.69 | 5.31 |
| 50 × 150mm | m | 0.19 | 3.04 | 2.32 | 0.80 | 6.16 |
| 75 × 125mm | m | 0.20 | 3.20 | 3.29 | 0.97 | 7.46 |
| 75 × 150mm | m | 0.23 | 3.70 | 3.80 | 1.12 | 8.62 |

Kerbs and bearers

| | | | | | | |
|---|---|---|---|---|---|---|
| 25 × 75mm | m | 0.13 | 2.08 | 0.85 | 0.44 | 3.37 |
| 25 × 100mm | m | 0.14 | 2.24 | 0.97 | 0.48 | 3.69 |
| 25 × 150mm | m | 0.17 | 2.72 | 1.35 | 0.61 | 4.68 |
| 38 × 75mm | m | 0.15 | 2.40 | 1.11 | 0.53 | 4.04 |
| 38 × 100mm | m | 0.16 | 2.56 | 1.48 | 0.61 | 4.65 |
| 38 × 150mm | m | 0.18 | 2.88 | 1.57 | 0.67 | 5.12 |
| 50 × 50mm | m | 0.16 | 2.56 | 1.16 | 0.56 | 4.28 |
| 50 × 75mm | m | 0.17 | 2.72 | 1.49 | 0.63 | 4.84 |
| 50 × 100mm | m | 0.18 | 2.88 | 1.56 | 0.67 | 5.11 |

| | Unit | Labour hours | Labour cost £ | Materials £ | O & P £ | Total £ |
|---|---|---|---|---|---|---|
| 75 × 75mm | m | 0.20 | 3.20 | 1.65 | 0.73 | 5.58 |
| 75 × 100mm | m | 0.22 | 3.52 | 2.22 | 0.86 | 6.60 |
| 75 × 125mm | m | 0.24 | 3.84 | 3.28 | 1.07 | 8.19 |

Solid strutting

| | | | | | | |
|---|---|---|---|---|---|---|
| 38 × 100mm | m | 0.40 | 6.40 | 1.49 | 1.18 | 9.07 |
| 50 × 100mm | m | 0.44 | 7.04 | 1.52 | 1.28 | 9.84 |
| 50 × 125mm | m | 0.48 | 7.68 | 1.72 | 1.41 | 10.81 |
| 50 × 150mm | m | 0.52 | 8.32 | 1.87 | 1.53 | 11.72 |

Herringbone strutting, 50 × 50mm
to joists, depth

| | | | | | | |
|---|---|---|---|---|---|---|
| 125mm | m | 0.60 | 9.60 | 1.60 | 1.68 | 12.88 |
| 150mm | m | 0.60 | 9.60 | 1.67 | 1.69 | 12.96 |
| 175mm | m | 0.60 | 9.60 | 1.72 | 1.70 | 13.02 |
| 240mm | m | 0.60 | 9.60 | 1.75 | 1.70 | 13.05 |

**Gutters and fascias**

Marine quality plywood in gutters
over 300mm wide, thickness

| | | | | | | |
|---|---|---|---|---|---|---|
| 12mm | m2 | 0.90 | 14.40 | 13.01 | 4.11 | 31.52 |
| 18mm | m2 | 1.00 | 16.00 | 17.88 | 5.08 | 38.96 |
| 25mm | m2 | 1.10 | 17.60 | 20.75 | 5.75 | 44.10 |

Marine quality plywood in gutters
150mm wide, thickness

| | | | | | | |
|---|---|---|---|---|---|---|
| 12mm | m | 0.28 | 4.48 | 2.07 | 0.98 | 7.53 |
| 18mm | m | 0.32 | 5.12 | 2.98 | 1.22 | 9.32 |
| 25mm | m | 0.36 | 5.76 | 3.46 | 1.38 | 10.60 |

| | Unit | Labour hours | Labour cost £ | Materials £ | O & P £ | Total £ |
|---|---|---|---|---|---|---|
| **Marine quality plywood in gutters 300mm wide, thickness** | | | | | | |
| 12mm | m | 0.36 | 5.76 | 4.06 | 1.47 | 11.29 |
| 18mm | m | 0.40 | 6.40 | 5.90 | 1.85 | 14.15 |
| 25mm | m | 0.44 | 7.04 | 6.74 | 2.07 | 15.85 |
| **Marine quality plywood in eaves, verges, soffits, fascias over 300mm wide, thickness** | | | | | | |
| 12mm | m2 | 0.80 | 12.80 | 13.01 | 3.87 | 29.68 |
| 18mm | m2 | 0.90 | 14.40 | 17.88 | 4.84 | 37.12 |
| 25mm | m2 | 1.00 | 16.00 | 19.36 | 5.30 | 40.66 |
| **Marine quality plywood in gutters 150mm wide, thickness** | | | | | | |
| 12mm | m | 0.24 | 3.84 | 2.17 | 0.90 | 6.91 |
| 18mm | m | 0.28 | 4.48 | 2.98 | 1.12 | 8.58 |
| 25mm | m | 0.32 | 5.12 | 3.46 | 1.29 | 9.87 |
| **Marine quality plywood in gutters 300mm wide, thickness** | | | | | | |
| 12mm | m | 0.32 | 5.12 | 4.06 | 1.38 | 10.56 |
| 18mm | m | 0.36 | 5.76 | 5.90 | 1.75 | 13.41 |
| 25mm | m | 0.40 | 6.40 | 6.74 | 1.97 | 15.11 |

## Wrought softwood supports

Bearers, framed

| | Unit | Labour hours | Labour cost £ | Materials £ | O & P £ | Total £ |
|---|---|---|---|---|---|---|
| 25 × 50mm | m | 0.09 | 1.44 | 0.72 | 0.32 | 2.48 |
| 38 × 50mm | m | 0.11 | 1.76 | 1.36 | 0.47 | 3.59 |
| 50 × 50mm | m | 0.12 | 1.92 | 1.35 | 0.49 | 3.76 |
| 50 × 75mm | m | 0.14 | 2.24 | 1.94 | 0.63 | 4.81 |
| 75 × 75mm | m | 0.16 | 2.56 | 3.84 | 0.96 | 7.36 |

| | Unit | Labour hours | Labour cost £ | Materials £ | O & P £ | Total £ |
|---|---|---|---|---|---|---|
| **Bearers, plugged and screwed** | | | | | | |
| 25 × 50mm | m | 0.14 | 2.24 | 0.72 | 0.44 | 3.40 |
| 38 × 50mm | m | 0.16 | 2.56 | 1.36 | 0.59 | 4.51 |
| 50 × 50mm | m | 0.17 | 2.72 | 1.35 | 0.61 | 4.68 |
| 50 × 75mm | m | 0.19 | 3.04 | 1.94 | 0.75 | 5.73 |
| 75 × 75mm | m | 0.21 | 3.36 | 3.84 | 1.08 | 8.28 |
| **Metal fixings** | | | | | | |
| Galvanised mild steel joist hangers, built in | | | | | | |
| 38 × 100mm | nr | 0.10 | 1.60 | 2.74 | 0.65 | 4.99 |
| 38 × 125mm | nr | 0.10 | 1.60 | 2.82 | 0.66 | 5.08 |
| 38 × 150mm | nr | 0.10 | 1.60 | 2.92 | 0.68 | 5.20 |
| 38 × 175mm | nr | 0.10 | 1.60 | 2.95 | 0.68 | 5.23 |
| 50 × 100mm | nr | 0.12 | 1.92 | 3.38 | 0.80 | 6.10 |
| 50 × 125mm | nr | 0.12 | 1.92 | 2.82 | 0.71 | 5.45 |
| 50 × 150mm | nr | 0.12 | 1.92 | 2.92 | 0.73 | 5.57 |
| 50 × 175mm | nr | 0.12 | 1.92 | 2.95 | 0.73 | 5.60 |
| 50 × 200mm | nr | 0.12 | 1.92 | 3.01 | 0.74 | 5.67 |
| 50 × 225mm | nr | 0.12 | 1.92 | 3.09 | 0.75 | 5.76 |
| 75 × 150mm | nr | 0.14 | 2.24 | 3.33 | 0.84 | 6.41 |
| 75 × 175mm | nr | 0.14 | 2.24 | 3.35 | 0.84 | 6.43 |
| 75 × 200mm | nr | 0.14 | 2.24 | 3.70 | 0.89 | 6.83 |
| 75 × 225mm | nr | 0.14 | 2.24 | 3.74 | 0.90 | 6.88 |
| Galvanised mild steel square toothed timber connectors, single sided, diameter | | | | | | |
| 38mm | nr | 0.03 | 0.48 | 0.26 | 0.11 | 0.85 |
| 50mm | nr | 0.03 | 0.48 | 0.27 | 0.11 | 0.86 |
| 63mm | nr | 0.03 | 0.48 | 0.29 | 0.12 | 0.89 |
| 75mm | nr | 0.03 | 0.48 | 0.31 | 0.12 | 0.91 |

| | Unit | Labour hours | Labour cost £ | Materials £ | O & P £ | Total £ |
|---|---|---|---|---|---|---|
| **Galvanised mild steel square toothed timber connectors, double sided, diameter** | | | | | | |
| 38mm | nr | 0.03 | 0.48 | 0.27 | 0.11 | 0.86 |
| 50mm | nr | 0.03 | 0.48 | 0.33 | 0.12 | 0.93 |
| 63mm | nr | 0.03 | 0.48 | 0.35 | 0.12 | 0.95 |
| 75mm | nr | 0.03 | 0.48 | 0.38 | 0.13 | 0.99 |
| **Galvanised mild steel restraint straps 30 × 2.5mm, girth** | | | | | | |
| 400mm | nr | 0.16 | 2.56 | 3.83 | 0.96 | 7.35 |
| 500mm | nr | 0.18 | 2.88 | 4.16 | 1.06 | 8.10 |
| 600mm | nr | 0.20 | 3.20 | 4.76 | 1.19 | 9.15 |
| 700mm | nr | 0.22 | 3.52 | 5.19 | 1.31 | 10.02 |
| 800mm | nr | 0.24 | 3.84 | 5.57 | 1.41 | 10.82 |
| 900mm | nr | 0.28 | 4.48 | 6.25 | 1.61 | 12.34 |
| 1000mm | nr | 0.30 | 4.80 | 6.56 | 1.70 | 13.06 |
| **Galvanised mild steel bolts, 100mm long, diameter** | | | | | | |
| 6mm | nr | 0.05 | 0.80 | 0.26 | 0.16 | 1.22 |
| 8mm | nr | 0.05 | 0.80 | 0.40 | 0.18 | 1.38 |
| 10mm | nr | 0.06 | 0.96 | 0.54 | 0.23 | 1.73 |
| 12mm | nr | 0.06 | 0.96 | 0.69 | 0.25 | 1.90 |
| 16mm | nr | 0.08 | 1.28 | 0.88 | 0.32 | 2.48 |
| 20mm | nr | 0.08 | 1.28 | 1.18 | 0.37 | 2.83 |
| **Galvanised mild steel bolts, 140mm long, diameter** | | | | | | |
| 6mm | nr | 0.05 | 0.80 | 0.56 | 0.20 | 1.56 |
| 8mm | nr | 0.05 | 0.80 | 0.69 | 0.22 | 1.71 |
| 10mm | nr | 0.06 | 0.96 | 0.82 | 0.27 | 2.05 |
| 12mm | nr | 0.06 | 0.96 | 0.98 | 0.29 | 2.23 |
| 16mm | nr | 0.08 | 1.28 | 1.14 | 0.36 | 2.78 |
| 20mm | nr | 0.08 | 1.28 | 1.39 | 0.40 | 3.07 |

| | Unit | Labour hours | Labour cost £ | Materials £ | O & P £ | Total £ |
|---|---|---|---|---|---|---|
| **Woodwool reinforced slabs 50mm thick, fixed to softwood joists in lengths** | | | | | | |
| 1800mm | m2 | 0.70 | 11.20 | 20.63 | 4.77 | 36.60 |
| 2100mm | m2 | 0.70 | 11.20 | 20.63 | 4.77 | 36.60 |
| 2400mm | m2 | 0.70 | 11.20 | 21.51 | 4.91 | 37.62 |
| 2700mm | m2 | 0.70 | 11.20 | 21.72 | 4.94 | 37.86 |
| 3000mm | m2 | 0.70 | 11.20 | 21.72 | 4.94 | 37.86 |
| **Woodwool reinforced slabs 75mm thick, fixed to softwood joists in lengths** | | | | | | |
| 1800mm | m2 | 0.75 | 12.00 | 29.05 | 6.16 | 47.21 |
| 2100mm | m2 | 0.75 | 12.00 | 29.05 | 6.16 | 47.21 |
| 2400mm | m2 | 0.75 | 12.00 | 30.19 | 6.33 | 48.52 |
| 2700mm | m2 | 0.75 | 12.00 | 30.62 | 6.39 | 49.01 |
| 3000mm | m2 | 0.75 | 12.00 | 30.62 | 6.39 | 49.01 |

**Flooring**

Wrought softwood boarding, butt-jointed

| | Unit | Labour hours | Labour cost £ | Materials £ | O & P £ | Total £ |
|---|---|---|---|---|---|---|
| 19 × 125mm | m2 | 0.60 | 9.60 | 8.60 | 2.73 | 20.93 |
| 25 × 100mm | m2 | 0.65 | 10.40 | 10.86 | 3.19 | 24.45 |
| 25 × 150mm | m2 | 0.60 | 9.60 | 11.73 | 3.20 | 24.53 |

Wrought softwood boarding, tongued and grooved

| | Unit | Labour hours | Labour cost £ | Materials £ | O & P £ | Total £ |
|---|---|---|---|---|---|---|
| 19 × 125mm | m2 | 0.70 | 11.20 | 9.23 | 3.06 | 23.49 |
| 25 × 100mm | m2 | 0.75 | 12.00 | 10.89 | 3.43 | 26.32 |
| 25 × 150mm | m2 | 0.70 | 11.20 | 12.39 | 3.54 | 27.13 |

| | Unit | Labour hours | Labour cost £ | Materials £ | O & P £ | Total £ |
|---|---|---|---|---|---|---|
| Chipboard flooring, butt-jointed, thickness | | | | | | |
| 18mm | m2 | 0.35 | 5.60 | 4.59 | 1.53 | 11.72 |
| 25mm | m2 | 0.40 | 6.40 | 5.53 | 1.79 | 13.72 |
| Chipboard flooring, tongued and grooved, thickness | | | | | | |
| 25mm | m2 | 0.50 | 8.00 | 5.98 | 2.10 | 16.08 |
| **Linings** | | | | | | |
| Melamine-faced chipboard, 15mm thick to walls | | | | | | |
| over 300mm wide | m2 | 1.10 | 17.60 | 5.61 | 3.48 | 26.69 |
| less than 300mm wide | m | 0.70 | 11.20 | 2.11 | 2.00 | 15.31 |
| Plain chipboard, 12mm thick to walls | | | | | | |
| over 300mm wide | m2 | 0.48 | 7.68 | 3.26 | 1.64 | 12.58 |
| less than 300mm wide | m | 0.28 | 4.48 | 1.03 | 0.83 | 6.34 |
| Birch-faced blockboard, 12mm thick to walls | | | | | | |
| over 300mm wide | m2 | 0.60 | 9.60 | 12.26 | 3.28 | 25.14 |
| less than 300mm wide | m | 0.40 | 6.40 | 4.08 | 1.57 | 12.05 |
| Standard quality hardboard, 3.2mm thick to walls | | | | | | |
| over 300mm wide | m2 | 0.40 | 6.40 | 2.50 | 1.34 | 10.24 |
| less than 300mm wide | m | 0.25 | 4.00 | 0.85 | 0.73 | 5.58 |

| | Unit | Labour hours | Labour cost £ | Materials £ | O & P £ | Total £ |
|---|---|---|---|---|---|---|
| **Standard quality hardboard, 6mm thick to walls** | | | | | | |
| over 300mm wide | m2 | 0.45 | 7.20 | 3.90 | 1.67 | 12.77 |
| less than 300mm wide | m | 0.30 | 4.80 | 1.34 | 0.92 | 7.06 |
| **Internal quality plywood, 4mm thick to walls** | | | | | | |
| over 300mm wide | m2 | 0.50 | 8.00 | 3.65 | 1.75 | 13.40 |
| less than 300mm wide | m | 0.30 | 4.80 | 1.50 | 0.95 | 7.25 |
| **Internal quality plywood, 6mm thick to walls** | | | | | | |
| over 300mm wide | m2 | 0.55 | 8.80 | 4.11 | 1.94 | 14.85 |
| less than 300mm wide | m | 0.34 | 5.44 | 1.68 | 1.07 | 8.19 |
| **Insulation board, 12mm thick to walls** | | | | | | |
| over 300mm wide | m2 | 0.40 | 6.40 | 2.52 | 1.34 | 10.26 |
| less than 300mm wide | m | 0.24 | 3.84 | 0.96 | 0.72 | 5.52 |

## Windows

| Standard softwood windows without glazing bars, type | Unit | Labour hours | Labour cost £ | Materials £ | O & P £ | Total £ |
|---|---|---|---|---|---|---|
| N07V, 488 × 750mm | nr | 0.75 | 12.00 | 98.82 | 16.62 | 127.44 |
| N09V, 488 × 900mm | nr | 1.00 | 16.00 | 103.72 | 17.96 | 137.68 |
| N12V, 488 × 1200mm | nr | 1.25 | 20.00 | 113.54 | 20.03 | 153.57 |
| 107C, 630 × 750mm | nr | 0.75 | 12.00 | 120.23 | 19.83 | 152.06 |
| 110C, 630 × 1050mm | nr | 1.00 | 16.00 | 125.66 | 21.25 | 162.91 |
| 112C, 630 × 1200mm | nr | 1.25 | 20.00 | 129.90 | 22.49 | 172.39 |
| 109V, 630 × 900mm | nr | 0.75 | 12.00 | 127.22 | 20.88 | 160.10 |
| 110V, 630 × 1050mm | nr | 1.25 | 20.00 | 132.10 | 22.82 | 174.92 |
| 112V, 630 × 1200mm | nr | 0.75 | 12.00 | 135.48 | 22.12 | 169.60 |
| 2N09W, 915 × 900mm | nr | 0.90 | 14.40 | 138.03 | 22.86 | 175.29 |

| | Unit | Labour hours | Labour cost £ | Materials £ | O & P £ | Total £ |
|---|---|---|---|---|---|---|
| 2N10W, 915 × 1050mm | nr | 1.15 | 18.40 | 143.50 | 24.29 | 186.19 |
| 2N12W, 915 × 1200mm | nr | 1.40 | 22.40 | 148.56 | 25.64 | 196.60 |
| 212W, 1200 × 1200mm | nr | 1.40 | 22.40 | 177.08 | 29.92 | 229.40 |
| 210C, 1200 × 1050mm | nr | 1.40 | 22.40 | 161.42 | 27.57 | 211.39 |
| 212T, 1200 × 1200mm | nr | 1.40 | 22.40 | 197.46 | 32.98 | 252.84 |

Standard hardwood windows
without glazing bars, type

| | Unit | Labour hours | Labour cost £ | Materials £ | O & P £ | Total £ |
|---|---|---|---|---|---|---|
| H2N10W, 915 × 1050mm | nr | 1.00 | 16.00 | 272.38 | 43.26 | 331.64 |
| H2N13W, 915 × 1050mm | nr | 1.25 | 20.00 | 287.03 | 46.05 | 353.08 |
| H2N15W, 915 × 1500mm | nr | 1.50 | 24.00 | 298.87 | 48.43 | 371.30 |
| H213W, 1200 × 1350mm | nr | 1.60 | 25.60 | 330.67 | 53.44 | 409.71 |
| H215W, 1200 × 1500mm | nr | 1.70 | 27.20 | 341.01 | 55.23 | 423.44 |
| H310CC, 1770 × 1050mm | nr | 2.00 | 32.00 | 502.20 | 80.13 | 614.33 |
| H312CC, 1770 × 1200mm | nr | 2.20 | 35.20 | 544.08 | 86.89 | 666.17 |

PVC-U windows, single glazed,
size

| | Unit | Labour hours | Labour cost £ | Materials £ | O & P £ | Total £ |
|---|---|---|---|---|---|---|
| 600 × 900mm | nr | 1.00 | 16.00 | 83.74 | 14.96 | 114.70 |
| 600 × 1200mm | nr | 1.30 | 20.80 | 159.47 | 27.04 | 207.31 |
| 1200 × 1200mm | nr | 1.60 | 25.60 | 199.38 | 33.75 | 258.73 |
| 1800 × 1200mm | nr | 2.00 | 32.00 | 227.03 | 38.85 | 297.88 |

PVC-U windows, double glazed,
size

| | Unit | Labour hours | Labour cost £ | Materials £ | O & P £ | Total £ |
|---|---|---|---|---|---|---|
| 600 × 900mm | nr | 1.00 | 16.00 | 108.47 | 18.67 | 143.14 |
| 600 × 1200mm | nr | 1.30 | 20.80 | 187.53 | 31.25 | 239.58 |
| 1200 × 1200mm | nr | 1.60 | 25.60 | 258.65 | 42.64 | 326.89 |
| 1800 × 1200mm | nr | 2.00 | 32.00 | 302.37 | 50.16 | 384.53 |

| | Unit | Labour hours | Labour cost £ | Materials £ | O & P £ | Total £ |
|---|---|---|---|---|---|---|

## Doors

Standard internal flush doors,
hardboard faced both sides,
35mm thick, size

| | | | | | | |
|---|---|---|---|---|---|---|
| 533 × 1981mm | nr | 1.20 | 19.20 | 28.65 | 7.18 | 55.03 |
| 610 × 1981mm | nr | 1.20 | 19.20 | 28.65 | 7.18 | 55.03 |
| 686 × 1981mm | nr | 1.20 | 19.20 | 28.65 | 7.18 | 55.03 |
| 762 × 1981mm | nr | 1.20 | 19.20 | 28.65 | 7.18 | 55.03 |

Standard internal flush doors,
hardboard faced both sides,
40mm thick, size

| | | | | | | |
|---|---|---|---|---|---|---|
| 533 × 1981mm | nr | 1.20 | 19.20 | 30.95 | 7.52 | 57.67 |
| 610 × 1981mm | nr | 1.20 | 19.20 | 30.95 | 7.52 | 57.67 |
| 686 × 1981mm | nr | 1.20 | 19.20 | 30.95 | 7.52 | 57.67 |
| 762 × 1981mm | nr | 1.20 | 19.20 | 30.95 | 7.52 | 57.67 |

Standard internal flush doors,
sapele faced both sides, 35mm
thick, size

| | | | | | | |
|---|---|---|---|---|---|---|
| 533 × 1981mm | nr | 1.30 | 20.80 | 43.28 | 9.61 | 73.69 |
| 610 × 1981mm | nr | 1.30 | 20.80 | 43.28 | 9.61 | 73.69 |
| 686 × 1981mm | nr | 1.30 | 20.80 | 43.28 | 9.61 | 73.69 |
| 762 × 1981mm | nr | 1.30 | 20.80 | 43.28 | 9.61 | 73.69 |

Standard internal flush doors,
sapele faced both sides, 40mm
thick, size

| | | | | | | |
|---|---|---|---|---|---|---|
| 533 × 1981mm | nr | 1.30 | 20.80 | 47.89 | 10.30 | 78.99 |
| 610 × 1981mm | nr | 1.30 | 20.80 | 47.89 | 10.30 | 78.99 |
| 686 × 1981mm | nr | 1.30 | 20.80 | 47.89 | 10.30 | 78.99 |
| 762 × 1981mm | nr | 1.30 | 20.80 | 47.89 | 10.30 | 78.99 |

| | Unit | Labour hours | Labour cost £ | Materials £ | O & P £ | Total £ |
|---|---|---|---|---|---|---|
| **Standard internal flush half-hour fire check doors, hardboard faced both sides, 44mm thick, size** | | | | | | |
| 686 × 1981mm | nr | 1.60 | 25.60 | 66.81 | 13.86 | 106.27 |
| 762 × 1981mm | nr | 1.60 | 25.60 | 68.15 | 14.06 | 107.81 |
| 726 × 2040mm | nr | 1.60 | 25.60 | 71.31 | 14.54 | 111.45 |
| 826 × 2040mm | nr | 1.60 | 25.60 | 71.92 | 14.63 | 112.15 |
| **Standard internal flush half-hour fire check doors, sapele faced both sides, 44mm thick, size** | | | | | | |
| 686 × 1981mm | nr | 1.80 | 28.80 | 91.84 | 18.10 | 138.74 |
| 762 × 1981mm | nr | 1.80 | 28.80 | 94.01 | 18.42 | 141.23 |
| 726 × 2040mm | nr | 1.80 | 28.80 | 94.28 | 18.46 | 141.54 |
| 826 × 2040mm | nr | 1.80 | 28.80 | 98.58 | 19.11 | 146.49 |
| **Standard external flush doors, plywood faced both sides, 54mm thick, size** | | | | | | |
| 762 × 1981mm | nr | 2.10 | 33.60 | 64.75 | 14.75 | 113.10 |
| 838 × 1981mm | nr | 2.10 | 33.60 | 67.53 | 15.17 | 116.30 |
| **Framed, ledged and braced doors, 44mm thick** | | | | | | |
| 726 × 2040mm | nr | 1.30 | 20.80 | 84.82 | 15.84 | 121.46 |
| 826 × 2040mm | nr | 1.30 | 20.80 | 86.87 | 16.15 | 123.82 |

## Door frames and linings

| | Unit | Labour hours | Labour cost £ | Materials £ | O & P £ | Total £ |
|---|---|---|---|---|---|---|
| **Standard door lining with loose stops, for door size** | | | | | | |
| 686 × 1981mm | | | | | | |
| 27 × 94mm | nr | 0.75 | 12.00 | 40.19 | 7.83 | 60.02 |
| 27 × 107mm | nr | 0.75 | 12.00 | 42.14 | 8.12 | 62.26 |

| | Unit | Labour hours | Labour cost £ | Materials £ | O & P £ | Total £ |
|---|---|---|---|---|---|---|
| 27 × 121mm | nr | 0.75 | 12.00 | 44.21 | 8.43 | 64.64 |
| 27 × 133mm | nr | 0.75 | 12.00 | 46.65 | 8.80 | 67.45 |
| 762 × 1981mm | | | | | | |
| 27 × 94mm | nr | 0.75 | 12.00 | 40.96 | 7.94 | 60.90 |
| 27 × 107mm | nr | 0.75 | 12.00 | 43.69 | 8.35 | 64.04 |
| 27 × 121mm | nr | 0.75 | 12.00 | 45.70 | 8.66 | 66.36 |
| 27 × 133mm | nr | 0.75 | 12.00 | 47.64 | 8.95 | 68.59 |
| 762 × 2040mm | | | | | | |
| 27 × 94mm | nr | 0.75 | 12.00 | 46.11 | 8.72 | 66.83 |
| 27 × 107mm | nr | 0.75 | 12.00 | 48.54 | 9.08 | 69.62 |
| 27 × 121mm | nr | 0.75 | 12.00 | 52.54 | 9.68 | 74.22 |
| 27 × 133mm | nr | 0.75 | 12.00 | 54.07 | 9.91 | 75.98 |

Standard rebated door frame,
for door size

| | Unit | Labour hours | Labour cost £ | Materials £ | O & P £ | Total £ |
|---|---|---|---|---|---|---|
| 686 × 1981mm | | | | | | |
| 35 × 107mm | nr | 0.75 | 12.00 | 42.14 | 8.12 | 62.26 |
| 35 × 133mm | nr | 0.75 | 12.00 | 42.14 | 8.12 | 62.26 |
| 762 × 1981mm | | | | | | |
| 35 × 107mm | nr | 0.75 | 12.00 | 42.14 | 8.12 | 62.26 |
| 35 × 133mm | nr | 0.75 | 12.00 | 42.14 | 8.12 | 62.26 |

Wrought softwood rebated
and rounded frames, size

| | Unit | Labour hours | Labour cost £ | Materials £ | O & P £ | Total £ |
|---|---|---|---|---|---|---|
| 38 × 100mm | m | 0.24 | 3.84 | 8.53 | 1.86 | 14.23 |
| 38 × 125mm | m | 0.24 | 3.84 | 11.10 | 2.24 | 17.18 |
| 38 × 140mm | m | 0.24 | 3.84 | 12.98 | 2.52 | 19.34 |
| 63 × 88mm | m | 0.26 | 4.16 | 13.62 | 2.67 | 20.45 |
| 63 × 100mm | m | 0.26 | 4.16 | 14.12 | 2.74 | 21.02 |
| 63 × 125mm | m | 0.26 | 4.16 | 15.88 | 3.01 | 23.05 |
| 75 × 100mm | m | 0.28 | 4.48 | 14.29 | 2.82 | 21.59 |
| 75 × 125mm | m | 0.28 | 4.48 | 20.68 | 3.77 | 28.93 |
| 75 × 140mm | m | 0.28 | 4.48 | 22.57 | 4.06 | 31.11 |

| | Unit | Labour hours | Labour cost £ | Materials £ | O & P £ | Total £ |
|---|---|---|---|---|---|---|
| **Mahogany rebated and rounded frames, size** | | | | | | |
| 38 × 100mm | m | 0.30 | 4.80 | 21.44 | 3.94 | 30.18 |
| 38 × 125mm | m | 0.30 | 4.80 | 23.22 | 4.20 | 32.22 |
| 38 × 140mm | m | 0.30 | 4.80 | 28.05 | 4.93 | 37.78 |
| 63 × 88mm | m | 0.32 | 5.12 | 30.75 | 5.38 | 41.25 |
| 63 × 100mm | m | 0.32 | 5.12 | 32.04 | 5.57 | 42.73 |
| 63 × 125mm | m | 0.32 | 5.12 | 35.82 | 6.14 | 47.08 |
| 75 × 100mm | m | 0.36 | 5.76 | 35.23 | 6.15 | 47.14 |
| 75 × 125mm | m | 0.36 | 5.76 | 47.44 | 7.98 | 61.18 |
| 75 × 140mm | m | 0.36 | 5.76 | 57.68 | 9.52 | 72.96 |

## Stairs

| | Unit | Labour hours | Labour cost £ | Materials £ | O & P £ | Total £ |
|---|---|---|---|---|---|---|
| **Wrought softwood open-tread staircase in one flight, 2600mm height, 2700mm going, 12 treads and 13 risers, width** | | | | | | |
| 850 mm | nr | 14.00 | 224.00 | 421.89 | 96.88 | 742.77 |
| 910 mm | nr | 14.00 | 224.00 | 446.93 | 100.64 | 771.57 |
| **Wrought softwood closed-tread staircase in one flight, 2600mm height, 2700mm going, 12 treads and 13 risers, width** | | | | | | |
| 850 mm | nr | 16.00 | 256.00 | 599.21 | 128.28 | 983.49 |
| 910 mm | nr | 16.00 | 256.00 | 634.28 | 133.54 | 1023.82 |
| **Hardwood open-tread staircase in one flight, 2600mm height, 2700mm going, 12 treads and 13 risers, width** | | | | | | |
| 850 mm | nr | 14.00 | 224.00 | 781.75 | 150.86 | 1156.61 |
| 910 mm | nr | 14.00 | 224.00 | 835.92 | 158.99 | 1218.91 |

| | Unit | Labour hours | Labour cost £ | Materials £ | O & P £ | Total £ |
|---|---|---|---|---|---|---|
| Hardwood closed-tread staircase in one flight, 2600mm height, 2700mm going, 12 treads and 13 risers, width | | | | | | |
| 850 mm | nr | 16.00 | 256.00 | 972.53 | 184.28 | 1412.81 |
| 910 mm | nr | 16.00 | 256.00 | 1020.42 | 191.46 | 1467.88 |

## Kitchen fittings

Fix only kitchen fittings

| | Unit | Labour hours | Labour cost £ | Materials £ | O & P £ | Total £ |
|---|---|---|---|---|---|---|
| wall units 200mm high, size | | | | | | |
| 300 × 600mm | nr | 1.00 | 16.00 | 0.00 | 2.40 | 18.40 |
| 300 × 1000mm | nr | 1.10 | 17.60 | 0.00 | 2.64 | 20.24 |
| 300 × 1200mm | nr | 1.20 | 19.20 | 0.00 | 2.88 | 22.08 |
| base units 750mm high, size | | | | | | |
| 600 × 900mm | nr | 1.20 | 19.20 | 0.00 | 2.88 | 22.08 |
| 600 × 1000mm | nr | 1.25 | 20.00 | 0.00 | 3.00 | 23.00 |
| 600 × 1200mm | nr | 1.30 | 20.80 | 0.00 | 3.12 | 23.92 |
| 900 × 900mm | nr | 1.30 | 20.80 | 0.00 | 3.12 | 23.92 |
| 900 × 1000mm | nr | 1.35 | 21.60 | 0.00 | 3.24 | 24.84 |
| 900 × 1200mm | nr | 1.40 | 22.40 | 0.00 | 3.36 | 25.76 |
| sink units 750mm high, size | | | | | | |
| 600 × 900mm | nr | 1.30 | 20.80 | 0.00 | 3.12 | 23.92 |
| 600 × 1000mm | nr | 1.35 | 21.60 | 0.00 | 3.24 | 24.84 |
| 600 × 1200mm | nr | 1.40 | 22.40 | 0.00 | 3.36 | 25.76 |
| worktops, size | | | | | | |
| 600 × 900mm | nr | 0.40 | 6.40 | 0.00 | 0.96 | 7.36 |
| 600 × 1000mm | nr | 0.50 | 8.00 | 0.00 | 1.20 | 9.20 |
| 600 × 1200mm | nr | 0.60 | 9.60 | 0.00 | 1.44 | 11.04 |

| | Unit | Labour hours | Labour cost £ | Materials £ | O & P £ | Total £ |
|---|---|---|---|---|---|---|

## Insulation

Glass fibre quilt laid over ceiling joists, thickness

| | | | | | | |
|---|---|---|---|---|---|---|
| 50mm | m2 | 0.15 | 2.40 | 3.80 | 0.93 | 7.13 |
| 80mm | m2 | 0.17 | 2.72 | 4.21 | 1.04 | 7.97 |
| 100mm | m2 | 0.20 | 3.20 | 4.66 | 1.18 | 9.04 |
| 150mm | m2 | 0.24 | 3.84 | 6.77 | 1.59 | 12.20 |

Glass fibre quilt fixed vertically to softwood, thickness

| | | | | | | |
|---|---|---|---|---|---|---|
| 50mm | m2 | 0.20 | 3.20 | 3.80 | 1.05 | 8.05 |
| 80mm | m2 | 0.23 | 3.68 | 4.21 | 1.18 | 9.07 |

Expanded polystyrene board fixed vertically to walls with adhesive, thickness

| | | | | | | |
|---|---|---|---|---|---|---|
| 12mm | m2 | 0.50 | 8.00 | 3.12 | 1.67 | 12.79 |
| 25mm | m2 | 0.60 | 9.60 | 6.02 | 2.34 | 17.96 |
| 50mm | m2 | 0.70 | 11.20 | 11.20 | 3.36 | 25.76 |

## Ironmongery

Fix only to softwood

door furniture

| | | | | | | |
|---|---|---|---|---|---|---|
| light butts | nr | 0.30 | 4.80 | 0.00 | 0.72 | 5.52 |
| medium butts | nr | 0.33 | 5.28 | 0.00 | 0.79 | 6.07 |
| heavy butts | nr | 0.35 | 5.60 | 0.00 | 0.84 | 6.44 |
| rising butts | nr | 0.40 | 6.40 | 0.00 | 0.96 | 7.36 |
| tee bands | nr | 1.00 | 16.00 | 0.00 | 2.40 | 18.40 |
| barrel bolts, small | nr | 0.55 | 8.80 | 0.00 | 1.32 | 10.12 |
| barrel bolts, medium | nr | 0.60 | 9.60 | 0.00 | 1.44 | 11.04 |
| barrel bolts, large | nr | 0.70 | 11.20 | 0.00 | 1.68 | 12.88 |
| tower bolts, small | nr | 0.55 | 8.80 | 0.00 | 1.32 | 10.12 |
| tower bolts, medium | nr | 0.60 | 9.60 | 0.00 | 1.44 | 11.04 |
| tower bolts, large | nr | 0.70 | 11.20 | 0.00 | 1.68 | 12.88 |

| | Unit | Labour hours | Labour cost £ | Materials £ | O & P £ | Total £ |
|---|---|---|---|---|---|---|
| panic bolts | nr | 2.20 | 35.20 | 0.00 | 5.28 | 40.48 |
| overhead door closer | nr | 1.00 | 16.00 | 0.00 | 2.40 | 18.40 |
| mortice latch | nr | 0.80 | 12.80 | 0.00 | 1.92 | 14.72 |
| cylinder night rim latch | nr | 1.00 | 16.00 | 0.00 | 2.40 | 18.40 |
| rim dead lock | nr | 0.70 | 11.20 | 0.00 | 1.68 | 12.88 |
| mortice dead lock | nr | 0.90 | 14.40 | 0.00 | 2.16 | 16.56 |
| mortice latch furniture | nr | 0.30 | 4.80 | 0.00 | 0.72 | 5.52 |
| Suffolk latch | nr | 0.65 | 10.40 | 0.00 | 1.56 | 11.96 |
| Norfolk latch | nr | 0.65 | 10.40 | 0.00 | 1.56 | 11.96 |
| postal knocker and plate | nr | 1.00 | 16.00 | 0.00 | 2.40 | 18.40 |
| pull handles | nr | 0.25 | 4.00 | 0.00 | 0.60 | 4.60 |
| push plates | nr | 0.25 | 4.00 | 0.00 | 0.60 | 4.60 |
| **window furniture** | | | | | | |
| casement stay | nr | 0.35 | 5.60 | 0.00 | 0.84 | 6.44 |
| casement fastener | nr | 0.25 | 4.00 | 0.00 | 0.60 | 4.60 |
| cockspur fastener | nr | 0.30 | 4.80 | 0.00 | 0.72 | 5.52 |
| **sundries** | | | | | | |
| numerals | nr | 0.05 | 0.80 | 0.00 | 0.12 | 0.92 |
| hat and coat hook | nr | 0.05 | 0.80 | 0.00 | 0.12 | 0.92 |
| toilet roll holder | nr | 0.10 | 1.60 | 0.00 | 0.24 | 1.84 |
| cabin hook | nr | 0.05 | 0.80 | 0.00 | 0.12 | 0.92 |
| **Fix only to hardwood** | | | | | | |
| **door furniture** | | | | | | |
| light butts | nr | 0.40 | 6.40 | 0.00 | 0.96 | 7.36 |
| medium butts | nr | 0.43 | 6.88 | 0.00 | 1.03 | 7.91 |
| heavy butts | nr | 0.45 | 7.20 | 0.00 | 1.08 | 8.28 |
| rising butts | nr | 0.50 | 8.00 | 0.00 | 1.20 | 9.20 |
| tee bands | nr | 1.15 | 18.40 | 0.00 | 2.76 | 21.16 |
| barrel bolts, small | nr | 0.75 | 12.00 | 0.00 | 1.80 | 13.80 |
| barrel bolts, medium | nr | 0.70 | 11.20 | 0.00 | 1.68 | 12.88 |
| barrel bolts, large | nr | 0.80 | 12.80 | 0.00 | 1.92 | 14.72 |
| tower bolts, small | nr | 0.65 | 10.40 | 0.00 | 1.56 | 11.96 |
| tower bolts, medium | nr | 0.70 | 11.20 | 0.00 | 1.68 | 12.88 |
| tower bolts, large | nr | 0.80 | 12.80 | 0.00 | 1.92 | 14.72 |
| panic bolts | nr | 2.40 | 38.40 | 0.00 | 5.76 | 44.16 |

| | Unit | Labour hours | Labour cost £ | Materials £ | O & P £ | Total £ |
|---|---|---|---|---|---|---|
| overhead door closer | nr | 1.15 | 18.40 | 0.00 | 2.76 | 21.16 |
| mortice latch | nr | 0.90 | 14.40 | 0.00 | 2.16 | 16.56 |
| cylinder night rim latch | nr | 1.15 | 18.40 | 0.00 | 2.76 | 21.16 |
| rim dead lock | nr | 0.80 | 12.80 | 0.00 | 1.92 | 14.72 |
| mortice dead lock | nr | 1.00 | 16.00 | 0.00 | 2.40 | 18.40 |
| mortice latch furniture | nr | 0.40 | 6.40 | 0.00 | 0.96 | 7.36 |
| Suffolk latch | nr | 0.75 | 12.00 | 0.00 | 1.80 | 13.80 |
| Norfolk latch | nr | 0.75 | 12.00 | 0.00 | 1.80 | 13.80 |
| postal knocker and plate | nr | 1.15 | 18.40 | 0.00 | 2.76 | 21.16 |
| pull handles | nr | 0.30 | 4.80 | 0.00 | 0.72 | 5.52 |
| push plates | nr | 0.30 | 4.80 | 0.00 | 0.72 | 5.52 |
| **window furniture** | | | | | | |
| casement stay | nr | 0.45 | 7.20 | 0.00 | 1.08 | 8.28 |
| casement fastener | nr | 0.35 | 5.60 | 0.00 | 0.84 | 6.44 |
| cockspur fastener | nr | 0.40 | 6.40 | 0.00 | 0.96 | 7.36 |
| **sundries** | | | | | | |
| numerals | nr | 0.07 | 1.12 | 0.00 | 0.17 | 1.29 |
| hat and coat hook | nr | 0.07 | 1.12 | 0.00 | 0.17 | 1.29 |
| toilet roll holder | nr | 0.14 | 2.24 | 0.00 | 0.34 | 2.58 |
| cabin hook | nr | 0.07 | 1.12 | 0.00 | 0.17 | 1.29 |

| | Unit | Labour hours | Labour cost £ | Materials £ | O & P £ | Total £ |
|---|---|---|---|---|---|---|
| **METALWORK** | | | | | | |
| **Balustrades** | | | | | | |
| Galvanised steel balustrades | | | | | | |
| 50 × 8mm flat bar | m | 0.50 | 8.00 | 28.22 | 5.43 | 41.65 |
| welded angles | nr | 0.75 | 12.00 | 0.00 | 1.80 | 13.80 |
| ramps | nr | 0.75 | 12.00 | 0.00 | 1.80 | 13.80 |
| flat bends | nr | 0.75 | 12.00 | 0.00 | 1.80 | 13.80 |
| 38 × 12mm rails | m | 0.50 | 8.00 | 41.02 | 7.35 | 56.37 |
| welded angles | nr | 0.75 | 12.00 | 0.00 | 1.80 | 13.80 |
| ramps | nr | 0.75 | 12.00 | 0.00 | 1.80 | 13.80 |
| flat bends | nr | 0.75 | 12.00 | 0.00 | 1.80 | 13.80 |
| **Lintels** | | | | | | |
| Standard galvanised steel lintels to 100mm thick wall, 143mm high | | | | | | |
| 900mm long | nr | 0.15 | 2.40 | 32.07 | 5.17 | 39.64 |
| 1050mm long | nr | 0.20 | 3.20 | 41.16 | 6.65 | 51.01 |
| 1200mm long | nr | 0.25 | 4.00 | 44.25 | 7.24 | 55.49 |
| Standard galvanised steel lintels to 256mm thick wall, 143mm high | | | | | | |
| 900mm long | nr | 0.20 | 3.20 | 39.17 | 6.36 | 48.73 |
| 1500mm long | nr | 0.25 | 4.00 | 60.47 | 9.67 | 74.14 |
| 2100mm long | nr | 0.30 | 4.80 | 87.19 | 13.80 | 105.79 |
| Standard galvanised steel lintels to 256mm thick wall, 219mm high | | | | | | |
| 2700mm long | nr | 0.75 | 12.00 | 141.49 | 23.02 | 176.51 |
| 4200mm long | nr | 1.00 | 16.00 | 292.96 | 46.34 | 355.30 |

|  | Unit | Labour hours | Labour cost £ | Materials £ | O & P £ | Total £ |
|---|---|---|---|---|---|---|
| **Sundries** | | | | | | |
| Galvanised mild steel water bars, size | | | | | | |
| 5 × 25mm | m | 0.50 | 8.00 | 3.96 | 1.79 | 13.75 |
| 5 × 30mm | m | 0.50 | 8.00 | 4.32 | 1.85 | 14.17 |
| 5 × 40mm | m | 0.50 | 8.00 | 5.67 | 2.05 | 15.72 |
| Galvanised steel dowels, size | | | | | | |
| 8mm diameter × 50mm | nr | 0.10 | 1.60 | 0.25 | 0.28 | 2.13 |
| 8mm diameter × 100mm | nr | 0.12 | 1.92 | 0.31 | 0.33 | 2.56 |
| 10mm diameter × 50mm | nr | 0.10 | 1.60 | 0.29 | 0.28 | 2.17 |
| 10mm diameter × 100mm | nr | 0.12 | 1.92 | 0.37 | 0.34 | 2.63 |
| 12mm diameter × 50mm | nr | 0.10 | 1.60 | 0.31 | 0.29 | 2.20 |
| 12mm diameter × 100mm | nr | 0.12 | 1.92 | 0.39 | 0.35 | 2.66 |

| | Unit | Labour hours | Labour cost £ | Materials £ | O & P £ | Total £ |
|---|---|---|---|---|---|---|

## ROOFING

### Tiling

Marley Marlden Plain granuled
or smooth finish clay tiles size
267 x 168mm, 65mm lap, 35°
pitch on reinforced underlay,
gauge 100mm, battens size

| | Unit | Labour hours | Labour cost £ | Materials £ | O & P £ | Total £ |
|---|---|---|---|---|---|---|
| 38 × 25mm | m2 | 1.44 | 23.04 | 35.45 | 8.77 | 67.26 |

Extra for

| | Unit | Labour hours | Labour cost £ | Materials £ | O & P £ | Total £ |
|---|---|---|---|---|---|---|
| nailing every tile with aluminium nails | m2 | 0.15 | 2.40 | 2.52 | 0.74 | 5.66 |
| cloak verge system | m | 0.25 | 4.00 | 11.49 | 2.32 | 17.81 |
| double course at eaves | m | 0.90 | 14.40 | 3.25 | 2.65 | 20.30 |
| segmental ridge tile | m | 0.60 | 9.60 | 14.57 | 3.63 | 27.80 |
| valley tiles | m | 0.60 | 9.60 | 12.12 | 3.26 | 24.98 |
| bonnet hip tiles | m | 0.60 | 9.60 | 7.60 | 2.58 | 19.78 |
| eaves vent system | m | 0.60 | 9.60 | 8.45 | 2.71 | 20.76 |
| ventilated ridge terminal | nr | 0.60 | 9.60 | 39.53 | 7.37 | 56.50 |
| vent terminal | nr | 0.60 | 9.60 | 59.37 | 10.35 | 79.31 |
| straight cutting | m | 0.20 | 3.20 | 0.00 | 0.48 | 3.68 |
| curved cutting | m | 0.30 | 4.80 | 0.00 | 0.72 | 5.52 |
| holes for pipes | nr | 0.40 | 6.40 | 0.00 | 0.96 | 7.36 |

Marley Ludlow Plus
concrete interlocking tiles size
387 × 229mm, battens size
38 ×25mm, reinforced underlay,
75mm lap, 22.5 to 45° pitch,

| | Unit | Labour hours | Labour cost £ | Materials £ | O & P £ | Total £ |
|---|---|---|---|---|---|---|
| battens size 38 × 25mm | m2 | 0.69 | 11.04 | 12.34 | 3.51 | 26.89 |

Extra for

| | Unit | Labour hours | Labour cost £ | Materials £ | O & P £ | Total £ |
|---|---|---|---|---|---|---|
| nailing every tile with aluminium nails | m2 | 0.03 | 0.48 | 0.63 | 0.17 | 1.28 |
| segmental ridge tile | m | 0.40 | 6.40 | 7.49 | 2.08 | 15.97 |
| segmental monoridge | m | 0.60 | 9.60 | 14.03 | 3.54 | 27.17 |

| | Unit | Labour hours | Labour cost £ | Materials £ | O & P £ | Total £ |
|---|---|---|---|---|---|---|
| dry ridge system | m | 0.60 | 9.60 | 13.71 | 3.50 | 26.81 |
| 1/3 hip riles | m | 0.40 | 6.40 | 7.67 | 2.11 | 16.18 |
| eaves vent system | m | 0.60 | 9.60 | 13.71 | 3.50 | 26.81 |
| ventilated ridge terminal | nr | 0.50 | 8.00 | 42.19 | 7.53 | 57.72 |
| gas vent terminal | nr | 0.50 | 8.00 | 59.95 | 10.19 | 78.14 |
| Marvent | nr | 0.50 | 8.00 | 20.42 | 4.26 | 32.68 |
| straight cutting | m | 0.20 | 3.20 | 0.00 | 0.48 | 3.68 |
| holes for pipes | nr | 0.40 | 6.40 | 0.00 | 0.96 | 7.36 |

Marley Modern smooth finish
concrete interlocking tiles size
420 × 330mm, battens size
38 ×25mm, reinforced underlay,
75mm lap, 22.5 to 45° pitch

| | Unit | Labour hours | Labour cost £ | Materials £ | O & P £ | Total £ |
|---|---|---|---|---|---|---|
| battens size 38 × 25mm | m2 | 0.59 | 9.44 | 19.50 | 4.34 | 33.28 |

Extra for

| | Unit | Labour hours | Labour cost £ | Materials £ | O & P £ | Total £ |
|---|---|---|---|---|---|---|
| nailing every tile with aluminium nails | m2 | 0.02 | 0.32 | 0.42 | 0.11 | 0.85 |
| verge, cloak system | m | 0.50 | 8.00 | 4.16 | 1.82 | 13.98 |
| dry verge system | m | 0.60 | 9.60 | 5.15 | 2.21 | 16.96 |
| Modern ridge tile | m | 0.40 | 6.40 | 7.81 | 2.13 | 16.34 |
| Modern monoridge | m | 0.60 | 9.60 | 13.94 | 3.53 | 27.07 |
| dry ridge system | m | 0.60 | 9.60 | 13.71 | 3.50 | 26.81 |
| 1/3 hip riles | m | 0.40 | 6.40 | 7.67 | 2.11 | 16.18 |
| eaves vent system | m | 0.60 | 9.60 | 13.71 | 3.50 | 26.81 |
| ventilated ridge terminal | nr | 0.50 | 8.00 | 42.19 | 7.53 | 57.72 |
| gas vent terminal | nr | 0.50 | 8.00 | 59.95 | 10.19 | 78.14 |
| Marvent | nr | 0.50 | 8.00 | 20.42 | 4.26 | 32.68 |
| straight cutting | m | 0.20 | 3.20 | 0.00 | 0.48 | 3.68 |
| holes for pipes | nr | 0.40 | 6.40 | 0.00 | 0.96 | 7.36 |

| | Unit | Labour hours | Labour cost £ | Materials £ | O & P £ | Total £ |
|---|---|---|---|---|---|---|
| Lafarge Norfolk concrete interlocking pantiles size 381 × 227mm, battens size 38 ×25mm, reinforced underlay, 75mm lap, battens size 38 × 25mm | m2 | 0.69 | 11.04 | 14.50 | 3.83 | 29.37 |
| Extra for | | | | | | |
| nailing every tile with aluminium nails | m2 | 0.02 | 0.32 | 0.42 | 0.11 | 0.85 |
| extra single course at | | | 0.00 | | | |
| verge | m | 0.40 | 6.40 | 6.69 | 1.96 | 15.05 |
| cloaked verge course | m | 0.40 | 6.40 | 8.26 | 2.20 | 16.86 |
| eaves vent system | m | 0.40 | 6.40 | 10.46 | 2.53 | 19.39 |
| universal ridge tile | m | 0.40 | 6.40 | 10.58 | 2.55 | 19.53 |
| universal hip tile | m | 0.40 | 6.40 | 10.58 | 2.55 | 19.53 |
| valley trough tile | m | 0.40 | 6.40 | 8.59 | 2.25 | 17.24 |
| universal gas flue ridge tile | nr | 0.50 | 8.00 | 55.50 | 9.53 | 73.03 |
| straight cutting | m | 0.20 | 3.20 | 0.00 | 0.48 | 3.68 |
| holes for pipes | nr | 0.40 | 6.40 | 0.00 | 0.96 | 7.36 |
| Lafarge Renown concrete interlocking tiles size 418 × 330mm, battens size 38 ×25mm, reinforced underlay, 75mm lap, battens size 38 × 25mm | m2 | 0.59 | 9.44 | 18.50 | 4.19 | 32.13 |
| Extra for | | | | | | |
| nailing every tile with aluminium nails | m2 | 0.03 | 0.48 | 0.63 | 0.17 | 1.28 |
| extra single course at | | | 0.00 | | | |
| verge | m | 0.40 | 6.40 | 6.69 | 1.96 | 15.05 |
| cloaked verge course | m | 0.40 | 6.40 | 8.26 | 2.20 | 16.86 |
| eaves vent system | m | 0.40 | 6.40 | 10.46 | 2.53 | 19.39 |
| universal ridge tile | m | 0.40 | 6.40 | 10.58 | 2.55 | 19.53 |
| universal hip tile | m | 0.40 | 6.40 | 10.58 | 2.55 | 19.53 |
| valley trough tile | m | 0.40 | 6.40 | 8.59 | 2.25 | 17.24 |

| | Unit | Labour hours | Labour cost £ | Materials £ | O & P £ | Total £ |
|---|---|---|---|---|---|---|
| universal gas flue ridge tile | nr | 0.50 | 8.00 | 55.50 | 9.53 | 73.03 |
| straight cutting | m | 0.20 | 3.20 | 0.00 | 0.48 | 3.68 |
| holes for pipes | nr | 0.40 | 6.40 | 0.00 | 0.96 | 7.36 |

**Fibre-cement slating**

Asbestos-free artificial slates
400 × 200mm, pitch 40-45°,
size 38 × 25mm softwood
battens, reinforced underlay

| | | | | | | |
|---|---|---|---|---|---|---|
| lap 90mm, gauge 155mm | m2 | 0.90 | 14.40 | 37.85 | 7.84 | 60.09 |

Asbestos-free artificial slates
500 × 250mm, pitch over 25°,
size 38 × 25mm softwood
battens, reinforced underlay

| | | | | | | |
|---|---|---|---|---|---|---|
| lap 90mm, gauge 205mm | m2 | 0.80 | 12.80 | 34.06 | 7.03 | 53.89 |

Asbestos-free artificial slates
600 × 300mm, pitch over 25°,
size 38 × 25mm softwood
battens, reinforced underlay

| | | | | | | |
|---|---|---|---|---|---|---|
| lap 100mm, gauge 250mm | m2 | 0.70 | 11.20 | 30.17 | 6.21 | 47.58 |

Extra for

| | | | | | | |
|---|---|---|---|---|---|---|
| verge, 150mm wide plain tile undercloak | m | 0.20 | 3.20 | 2.22 | 0.81 | 6.23 |
| double course at eaves | m | 0.35 | 5.60 | 6.99 | 1.89 | 14.48 |
| half-round ridge tile | m | 0.40 | 6.40 | 25.20 | 4.74 | 36.34 |
| valley tiles | m | 0.60 | 9.60 | 25.20 | 5.22 | 40.02 |
| straight cutting | m | 0.20 | 3.20 | 0.00 | 0.48 | 3.68 |
| holes for pipes | nr | 0.40 | 6.40 | 0.00 | 0.96 | 7.36 |

| | Unit | Labour hours | Labour cost £ | Materials £ | O & P £ | Total £ |
|---|---|---|---|---|---|---|
| **Natural slating** | | | | | | |
| Blue/grey Welsh slates size 405 × 205mm, 75mm lap, 50 × 25mm softwood battens, reinforced underlay | | | | | | |
| sloping | m2 | 1.20 | 19.20 | 56.05 | 11.77 | 90.22 |
| vertical | m2 | 1.40 | 22.40 | 56.05 | 8.41 | 64.46 |
| Extra for | | | | | | |
| double eaves course | m | 0.50 | 8.00 | 26.26 | 5.62 | 43.08 |
| verge undercloak course | m | 0.70 | 11.20 | 19.66 | 4.63 | 35.49 |
| angled ridge or hip tiles | m | 0.70 | 11.20 | 17.05 | 2.56 | 19.61 |
| mitred hips including cutting both sides | m | 0.70 | 11.20 | 28.91 | 5.78 | 44.29 |
| straight cutting | m | 0.60 | 9.60 | 0.00 | 0.96 | 7.36 |
| holes for pipes | nr | 0.40 | 6.40 | 0.00 | 1.20 | 9.20 |
| fix only lead soakers | nr | 0.50 | 8.00 | 0.00 | 0.00 | 0.00 |
| Blue/grey Welsh slates size 510 × 255mm, 75mm lap, 50 × 25mm softwood battens, reinforced underlay | | | | | | |
| sloping | m2 | 1.00 | 16.00 | 58.41 | 11.64 | 89.25 |
| vertical | m2 | 1.20 | 19.20 | 58.41 | 8.76 | 67.17 |
| Extra for | | | | | | |
| double eaves course | m | 0.50 | 8.00 | 32.86 | 6.61 | 50.67 |
| verge undercloak course | m | 0.70 | 11.20 | 20.86 | 4.81 | 36.87 |
| angled ridge or hip tiles | m | 0.70 | 11.20 | 17.05 | 2.56 | 19.61 |
| mitred hips including cutting both sides | m | 0.70 | 11.20 | 28.91 | 5.78 | 44.29 |
| straight cutting | m | 0.60 | 9.60 | 0.00 | 0.96 | 7.36 |
| holes for pipes | nr | 0.40 | 6.40 | 0.00 | 1.20 | 9.20 |
| fix only lead soakers | nr | 0.50 | 8.00 | 0.00 | 0.00 | 0.00 |

| | Unit | Labour hours | Labour cost £ | Materials £ | O & P £ | Total £ |
|---|---|---|---|---|---|---|
| **Blue/grey Welsh slates size 610 × 305mm, 75mm lap, 50 × 25mm softwood battens, reinforced underlay** | | | | | | |
| sloping | m2 | 0.80 | 12.80 | 64.20 | 11.55 | 88.55 |
| vertical | m2 | 1.00 | 16.00 | 64.20 | 12.03 | 92.23 |
| **Extra for** | | | | | | |
| double eaves course | m | 0.50 | 8.00 | 37.07 | 6.76 | 51.83 |
| verge undercloak course | m | 0.70 | 11.20 | 22.87 | 5.11 | 39.18 |
| angled ridge or hip tiles | m | 0.70 | 11.20 | 17.05 | 4.24 | 32.49 |
| mitred hips including cutting | | | 0.00 | | | |
| both sides | m | 0.70 | 11.20 | 28.91 | 6.02 | 46.13 |
| straight cutting | m | 0.60 | 9.60 | 0.00 | 1.44 | 11.04 |
| holes for pipes | nr | 0.40 | 6.40 | 0.00 | 0.96 | 7.36 |
| fix only lead soakers | nr | 0.50 | 8.00 | 0.00 | 1.20 | 9.20 |
| **Westmorland green slates in random lengths 450 - 230mm, 75mm laps, 50 × 25mm softwood battens, reinforced underlay** | | | | | | |
| sloping | m2 | 1.35 | 21.60 | 159.75 | 27.20 | 208.55 |
| vertical | m2 | 1.55 | 24.80 | 159.75 | 27.68 | 212.23 |
| **Extra for** | | | | | | |
| double eaves course | m | 0.50 | 8.00 | 37.07 | 6.76 | 51.83 |
| verge undercloak course | m | 0.70 | 11.20 | 22.87 | 5.11 | 39.18 |
| angled ridge or hip tiles | m | 0.70 | 11.20 | 17.05 | 4.24 | 32.49 |
| straight cutting | m | 0.60 | 9.60 | 0.00 | 1.44 | 11.04 |
| holes for pipes | nr | 0.40 | 6.40 | 0.00 | 0.96 | 7.36 |
| fix only lead soakers | nr | 0.50 | 8.00 | 0.00 | 1.20 | 9.20 |

| | Unit | Labour hours | Labour cost £ | Materials £ | O & P £ | Total £ |
|---|---|---|---|---|---|---|

### Reconstructed stone slating

Marley Monarch interlocking slate
size 325 × 330mm, 38 × 25mm
softwood battens, reinforced
underlay

| | Unit | Labour hours | Labour cost £ | Materials £ | O & P £ | Total £ |
|---|---|---|---|---|---|---|
| 75mm lap, pitch 25-90° | m2 | 0.60 | 9.60 | 28.34 | 5.69 | 43.63 |
| 100mm lap, pitch 25-90° | m2 | 0.65 | 10.40 | 28.34 | 5.81 | 44.55 |

Extra for

| | Unit | Labour hours | Labour cost £ | Materials £ | O & P £ | Total £ |
|---|---|---|---|---|---|---|
| nailing every tile with aluminium nails | m2 | 0.02 | 0.32 | 0.42 | 0.11 | 0.85 |
| interlocking dry verge | m | 0.60 | 9.60 | 5.15 | 2.21 | 16.96 |
| eaves vent system | m | 0.60 | 9.60 | 13.71 | 3.50 | 26.81 |
| Modern ridge tiles | m | 0.40 | 6.40 | 7.81 | 2.13 | 16.34 |
| Modern hip tiles | m | 0.40 | 6.40 | 7.81 | 2.13 | 16.34 |
| segmental ridge tile | m | 0.40 | 6.40 | 7.56 | 2.09 | 16.05 |
| Modern monoridge | m | 0.60 | 9.60 | 13.94 | 3.53 | 27.07 |
| dry ridge system | m | 0.60 | 9.60 | 13.71 | 3.50 | 26.81 |
| segmental hip tiles | m | 0.40 | 6.40 | 7.32 | 2.06 | 15.78 |
| ventilated ridge terminal | nr | 0.50 | 8.00 | 42.19 | 7.53 | 57.72 |
| gas vent terminal | nr | 0.50 | 8.00 | 59.95 | 10.19 | 78.14 |
| straight cutting | m | 0.20 | 3.20 | 0.00 | 0.48 | 3.68 |
| holes for pipes | nr | 0.40 | 6.40 | 0.00 | 0.96 | 7.36 |

Lafarge Cambrian through-
coloured slates size 300 × 336mm,
38 × 25mm softwood battens,

| | Unit | Labour hours | Labour cost £ | Materials £ | O & P £ | Total £ |
|---|---|---|---|---|---|---|
| reinforced underlay | m2 | 0.60 | 9.60 | 28.49 | 5.71 | 43.80 |

Extra for

| | Unit | Labour hours | Labour cost £ | Materials £ | O & P £ | Total £ |
|---|---|---|---|---|---|---|
| nailing every tile with aluminium nails | m2 | 0.02 | 0.32 | 0.45 | 0.12 | 0.89 |
| dryvent ridge | m | 0.75 | 12.00 | 18.70 | 4.61 | 35.31 |
| half-round ridge or hip tile | m | 0.40 | 6.40 | 11.21 | 2.64 | 20.25 |
| verge system | m | 0.40 | 6.40 | 12.38 | 2.82 | 21.60 |
| eaves vent system | m | 0.60 | 9.60 | 10.89 | 3.07 | 23.56 |

| | Unit | Labour hours | Labour cost £ | Materials £ | O & P £ | Total £ |
|---|---|---|---|---|---|---|
| gas flue ridge terminal | nr | 0.50 | 8.00 | 58.83 | 10.02 | 76.85 |
| valley trough tiles | m | 0.40 | 6.40 | 9.10 | 2.33 | 17.83 |
| straight cutting | m | 0.20 | 3.20 | 0.00 | 0.48 | 3.68 |
| holes for pipes | nr | 0.40 | 6.40 | 0.00 | 0.96 | 7.36 |

**Lead sheet coverings**

Milled lead roof coverings to
roof with falls less than 10°

| | | | | | | |
|---|---|---|---|---|---|---|
| code 5 | m2 | 3.10 | 49.60 | 33.98 | 12.54 | 96.12 |
| code 6 | m2 | 3.20 | 51.20 | 37.51 | 13.31 | 102.02 |

Milled lead roof coverings to
dormers

| | | | | | | |
|---|---|---|---|---|---|---|
| code 5 | m2 | 3.80 | 60.80 | 33.98 | 14.22 | 109.00 |
| code 6 | m2 | 4.00 | 64.00 | 37.51 | 15.23 | 116.74 |

Lead flashings

| | | | | | | |
|---|---|---|---|---|---|---|
| code 5, horizontal | | | | | | |
| 150mm girth | m | 0.60 | 9.60 | 5.10 | 2.21 | 16.91 |
| 200mm girth | m | 0.70 | 11.20 | 8.15 | 2.90 | 22.25 |
| 300mm girth | m | 0.85 | 13.60 | 10.21 | 3.57 | 27.38 |
| code 5, sloping | | | | | | |
| 150mm girth | m | 0.80 | 12.80 | 5.10 | 2.69 | 20.59 |
| 200mm girth | m | 0.85 | 13.60 | 8.15 | 3.26 | 25.01 |
| 300mm girth | m | 0.90 | 14.40 | 10.21 | 3.69 | 28.30 |

Lead aprons

| | | | | | | |
|---|---|---|---|---|---|---|
| code 5, horizontal | | | | | | |
| 200mm girth | m | 0.90 | 14.40 | 8.15 | 3.38 | 25.93 |
| 300mm girth | m | 0.95 | 15.20 | 10.21 | 3.81 | 29.22 |
| 400mm girth | m | 1.00 | 16.00 | 15.30 | 4.70 | 36.00 |

| | Unit | Labour hours | Labour cost £ | Materials £ | O & P £ | Total £ |
|---|---|---|---|---|---|---|
| code 5, sloping | | | | | | |
| 200mm girth | m | 0.95 | 15.20 | 8.15 | 3.50 | 26.85 |
| 300mm girth | m | 1.00 | 16.00 | 10.21 | 3.93 | 30.14 |
| 400mm girth | m | 1.05 | 16.80 | 15.30 | 4.82 | 36.92 |
| **Lead sills** | | | | | | |
| code 5, horizontal | | | | | | |
| 200mm girth | m | 0.80 | 12.80 | 8.15 | 3.14 | 24.09 |
| 300mm girth | m | 0.90 | 14.40 | 10.21 | 3.69 | 28.30 |
| 400mm girth | m | 1.00 | 16.00 | 15.30 | 4.70 | 36.00 |
| code 5, sloping | | | | | | |
| 200mm girth | m | 0.90 | 14.40 | 8.15 | 3.38 | 25.93 |
| 300mm girth | m | 1.00 | 16.00 | 10.21 | 3.93 | 30.14 |
| 400mm girth | m | 1.10 | 17.60 | 15.30 | 4.94 | 37.84 |
| **Lead hips** | | | | | | |
| code 5, sloping | | | | | | |
| 150mm girth | m | 0.95 | 15.20 | 8.15 | 3.50 | 26.85 |
| 200mm girth | m | 1.15 | 18.40 | 10.21 | 4.29 | 32.90 |
| 300mm girth | m | 1.35 | 21.60 | 15.30 | 5.54 | 42.44 |
| **Lead kerbs** | | | | | | |
| code 5, sloping | | | | | | |
| 300mm girth | m | 1.00 | 16.00 | 10.21 | 3.93 | 30.14 |
| 400mm girth | m | 1.10 | 17.60 | 15.30 | 4.94 | 37.84 |
| **Lead valleys** | | | | | | |
| code 5, sloping | | | | | | |
| 400mm girth | m | 1.10 | 17.60 | 15.30 | 4.94 | 37.84 |
| 600mm girth | m | 1.25 | 20.00 | 20.24 | 6.04 | 46.28 |

| | Unit | Labour hours | Labour cost £ | Materials £ | O & P £ | Total £ |
|---|---|---|---|---|---|---|
| **Lead gutters** | | | | | | |
| code 5, sloping | | | | | | |
| 600mm girth | m | 1.25 | 20.00 | 20.24 | 6.04 | 46.28 |
| 800mm girth | m | 1.55 | 24.80 | 26.49 | 7.69 | 58.98 |
| **Lead slates** | | | | | | |
| code 5 | | | | | | |
| size 400 × 400mm with 200mm high collar 100mm diameter | nr | 1.50 | 24.00 | 13.97 | 5.70 | 43.67 |
| size 400 × 400mm with 200mm high collar 150mm diameter | nr | 1.70 | 27.20 | 14.06 | 6.19 | 47.45 |
| **Copper sheet coverings** | | | | | | |
| Copper sheet coverings to roof with falls less than 10° | | | | | | |
| 0.55mm thick | m2 | 3.20 | 51.20 | 36.49 | 13.15 | 100.84 |
| 0.61mm thick | m2 | 3.20 | 51.20 | 38.44 | 13.45 | 103.09 |
| Copper sheet coverings to roof with falls more than 10° but less than 50° | | | | | | |
| 0.55mm thick | m2 | 3.50 | 56.00 | 36.49 | 13.87 | 106.36 |
| 0.61mm thick | m2 | 3.50 | 56.00 | 38.44 | 14.17 | 108.61 |
| Copper flashings | | | | | | |
| 0.55mm thick, horizontal | | | | | | |
| 150mm girth | m | 0.55 | 8.80 | 5.46 | 2.14 | 16.40 |
| 200mm girth | m | 0.65 | 10.40 | 7.16 | 2.63 | 20.19 |
| 300mm girth | m | 0.75 | 12.00 | 10.94 | 3.44 | 26.38 |

| | Unit | Labour hours | Labour cost £ | Materials £ | O & P £ | Total £ |
|---|---|---|---|---|---|---|
| **0.55mm thick, sloping** | | | | | | |
| 150mm girth | m | 0.60 | 9.60 | 5.46 | 2.26 | 17.32 |
| 200mm girth | m | 0.70 | 11.20 | 7.29 | 2.77 | 21.26 |
| 300mm girth | m | 0.80 | 12.80 | 10.94 | 3.56 | 27.30 |
| **0.61mm thick, horizontal** | | | | | | |
| 150mm girth | m | 0.55 | 8.80 | 5.77 | 2.19 | 16.76 |
| 200mm girth | m | 0.65 | 10.40 | 7.26 | 2.65 | 20.31 |
| 300mm girth | m | 0.75 | 12.00 | 11.53 | 3.53 | 27.06 |
| **0.61mm thick, sloping** | | | | | | |
| 150mm girth | m | 0.60 | 9.60 | 5.77 | 2.31 | 17.68 |
| 200mm girth | m | 0.70 | 11.20 | 7.26 | 2.77 | 21.23 |
| 300mm girth | m | 0.80 | 12.80 | 11.53 | 3.65 | 27.98 |
| **Copper soakers size** | | | | | | |
| 175 × 175mm | nr | 0.25 | 4.00 | 2.39 | 0.96 | 7.35 |
| 300 × 175mm | nr | 0.25 | 4.00 | 2.83 | 1.02 | 7.85 |

**Built-up roofing**

Two layer built-up bituminous
felt roof coverings, the layers
bonded with hot bitumen

| | Unit | Labour hours | Labour cost £ | Materials £ | O & P £ | Total £ |
|---|---|---|---|---|---|---|
| laid flat | m2 | 0.35 | 5.60 | 7.65 | 1.99 | 15.24 |
| laid sloping | m2 | 0.40 | 6.40 | 7.65 | 2.11 | 16.16 |

Extra for

| | Unit | Labour hours | Labour cost £ | Materials £ | O & P £ | Total £ |
|---|---|---|---|---|---|---|
| **aprons** | | | | | | |
| 100mm girth | m | 0.25 | 4.00 | 1.79 | 0.87 | 6.66 |
| 150mm girth | m | 0.30 | 4.80 | 2.57 | 1.11 | 8.48 |
| **skirtings** | | | | | | |
| 100mm girth | m | 0.25 | 4.00 | 1.79 | 0.87 | 6.66 |
| 150mm girth | m | 0.30 | 4.80 | 2.57 | 1.11 | 8.48 |

| | Unit | Labour hours | Labour cost £ | Materials £ | O & P £ | Total £ |
|---|---|---|---|---|---|---|
| flashings | | | | | | |
| 100mm girth | m | 0.25 | 4.00 | 1.66 | 0.85 | 6.51 |
| 150mm girth | m | 0.30 | 4.80 | 2.39 | 1.08 | 8.27 |
| Form collars around 100mm diameter pipe | nr | 1.25 | 20.00 | 3.15 | 3.47 | 26.62 |
| Form collars around 150mm diameter pipe | nr | 1.35 | 21.60 | 5.72 | 4.10 | 31.42 |
| Three layer built-up bituminous felt roof coverings, the layers bonded with hot bitumen | | | | | | |
| laid flat | m2 | 0.50 | 8.00 | 11.86 | 2.98 | 22.84 |
| laid sloping | m2 | 0.55 | 8.80 | 11.86 | 3.10 | 23.76 |
| Extra for | | | | | | |
| aprons | | | | | | |
| 100mm girth | m | 0.30 | 4.80 | 2.11 | 1.04 | 7.95 |
| 150mm girth | m | 0.35 | 5.60 | 2.82 | 1.26 | 9.68 |
| skirtings | | | | | | |
| 100mm girth | m | 0.30 | 4.80 | 2.11 | 1.04 | 7.95 |
| 150mm girth | m | 0.35 | 5.60 | 2.82 | 1.26 | 9.68 |
| flashings | | | | | | |
| 100mm girth | m | 0.30 | 4.80 | 2.11 | 1.04 | 7.95 |
| 150mm girth | m | 0.35 | 5.60 | 2.82 | 1.26 | 9.68 |
| Form collars around 100mm diameter pipe | nr | 1.50 | 24.00 | 5.07 | 4.36 | 33.43 |
| Form collars around 150mm diameter pipe | nr | 1.60 | 25.60 | 5.72 | 4.70 | 36.02 |

| | Unit | Specialist price £ | O & P £ | Total £ |
|---|---|---|---|---|

## ASPHALT WORK

The following are specialist sub-
contractor's prices

### Damp proofing and tanking to BS 1097

20mm two coat mastic asphalt
coverings to concrete surfaces

| | Unit | Specialist price £ | O & P £ | Total £ |
|---|---|---|---|---|
| flat | m2 | 37.58 | 5.64 | 43.22 |
| sloping 10-45° | m2 | 41.15 | 6.17 | 47.32 |
| sloping 46-90° | m2 | 44.94 | 6.74 | 51.68 |
| vertical | m2 | 44.08 | 6.61 | 50.69 |

20mm two coat mastic asphalt
coverings to concrete surfaces
not exceeding 150mm wide

| | Unit | Specialist price £ | O & P £ | Total £ |
|---|---|---|---|---|
| flat | m | 5.69 | 0.85 | 6.54 |
| sloping 10-45° | m | 6.26 | 0.94 | 7.20 |
| sloping 46-90° | m | 6.94 | 1.04 | 7.98 |
| vertical | m | 6.94 | 1.04 | 7.98 |

20mm two coat mastic asphalt
coverings to concrete surfaces
150-300mm wide

| | Unit | Specialist price £ | O & P £ | Total £ |
|---|---|---|---|---|
| flat | m | 11.27 | 1.69 | 12.96 |
| sloping 10-45° | m | 12.42 | 1.86 | 14.28 |
| sloping 46-90° | m | 13.55 | 2.03 | 15.58 |
| vertical | m | 13.55 | 2.03 | 15.58 |

30mm three coat mastic asphalt
coverings to concrete surfaces

| | Unit | Specialist price £ | O & P £ | Total £ |
|---|---|---|---|---|
| flat | m2 | 67.43 | 10.11 | 77.54 |
| sloping 10-45 ° | m2 | 71.08 | 10.66 | 81.74 |

| | Unit | Specialist price £ | O & P £ | Total £ |
|---|---|---|---|---|
| sloping 46-90° | m2 | 75.70 | 11.36 | 87.06 |
| vertical | m2 | 75.70 | 11.36 | 87.06 |

30mm three coat mastic asphalt
coverings to concrete surfaces
not exceeding 150mm wide

| | Unit | Specialist price £ | O & P £ | Total £ |
|---|---|---|---|---|
| flat | m | 10.12 | 1.52 | 11.64 |
| sloping 10-45° | m | 11.07 | 1.66 | 12.73 |
| sloping 46-90° | m | 12.27 | 1.84 | 14.11 |
| vertical | m | 12.28 | 1.84 | 14.12 |

30mm three coat mastic asphalt
coverings to concrete surfaces
150-300mm wide

| | Unit | Specialist price £ | O & P £ | Total £ |
|---|---|---|---|---|
| flat | m | 20.33 | 3.05 | 23.38 |
| sloping 10-45° | m | 22.59 | 3.39 | 25.98 |
| sloping 46-90° | m | 24.63 | 3.69 | 28.32 |
| vertical | m | 24.63 | 3.69 | 28.32 |

**Flooring to BS 1076**

20mm two coat mastic asphalt
coverings to flat concrete surfaces

| | Unit | Specialist price £ | O & P £ | Total £ |
|---|---|---|---|---|
| exceeding 300mm | m2 | 30.92 | 4.64 | 35.56 |
| not exceeding 150mm wide | m | 4.70 | 0.71 | 5.41 |
| 150-300mm wide | m | 9.50 | 1.43 | 10.93 |

20mm two coat mastic asphalt
coverings to sloping concrete
surfaces

| | Unit | Specialist price £ | O & P £ | Total £ |
|---|---|---|---|---|
| exceeding 300mm | m2 | 34.31 | 5.15 | 39.46 |
| not exceeding 150mm wide | m | 5.48 | 0.82 | 6.30 |
| 150-300mm wide | m | 10.81 | 1.62 | 12.43 |

| | Unit | Specialist price £ | O & P £ | Total £ |
|---|---|---|---|---|
| **Extra for** | | | | |
| working into recessed frames | m | 2.73 | 0.41 | 3.14 |
| working into channels 300mm girth | m | 5.34 | 0.80 | 6.14 |
| working into channels 600mm girth | m | 10.90 | 1.64 | 12.54 |
| **30mm three coat mastic asphalt coverings to flat concrete surfaces** | | | | |
| exceeding 300mm | m2 | 55.19 | 8.28 | 63.47 |
| not exceeding 150mm wide | m | 8.74 | 1.31 | 10.05 |
| 150-300mm wide | m | 17.54 | 2.63 | 20.17 |
| **30mm three coat mastic asphalt coverings to sloping concrete surfaces** | | | | |
| exceeding 300mm | m2 | 60.28 | 9.04 | 69.32 |
| not exceeding 150mm wide | m | 9.20 | 1.38 | 10.58 |
| 150-300mm wide | m | 18.14 | 2.72 | 20.86 |
| **Extra for** | | | | |
| working into recessed frames | m | 4.13 | 0.62 | 4.75 |
| working into channels 300mm girth | m | 9.65 | 1.45 | 11.10 |
| working into channels 600mm girth | m | 16.93 | 2.54 | 19.47 |

**Roofing to BS 988**

| 20mm two coat mastic asphalt coverings | Unit | Specialist price £ | O & P £ | Total £ |
|---|---|---|---|---|
| flat | m2 | 40.07 | 6.01 | 46.08 |
| sloping 10-45° | m2 | 42.30 | 6.35 | 48.65 |
| sloping 46-90° | m2 | 46.46 | 6.97 | 53.43 |
| vertical | m2 | 46.46 | 6.97 | 53.43 |

| | Unit | Specialist price £ | O & P £ | Total £ |
|---|---|---|---|---|
| **20mm two coat mastic asphalt coverings to concrete surfaces not exceeding 150mm wide** | | | | |
| flat | m | 6.03 | 0.90 | 6.93 |
| sloping 10-45° | m | 6.33 | 0.95 | 7.28 |
| sloping 46-90° | m | 7.05 | 1.06 | 8.11 |
| vertical | m | 7.05 | 1.06 | 8.11 |
| **20mm two coat mastic asphalt coverings to concrete surfaces 150-300mm wide** | | | | |
| flat | m | 12.12 | 1.82 | 13.94 |
| sloping 10-45° | m | 12.73 | 1.91 | 14.64 |
| sloping 46-90° | m | 13.97 | 2.10 | 16.07 |
| vertical | m | 13.97 | 2.10 | 16.07 |

| | Unit | Labour hours | Labour cost £ | Materials £ | O & P £ | Total £ |
|---|---|---|---|---|---|---|

## FLOOR, WALL AND CEILING FINISHINGS

### Screeds

Cement and sand (1:3) screed in beds with steel trowelled finish to level floors or to falls not exceeding 15 degrees from the horizontal

| | Unit | Labour hours | Labour cost £ | Materials £ | O & P £ | Total £ |
|---|---|---|---|---|---|---|
| 25mm thick | | | | | | |
|   over 300mm wide | m2 | 0.40 | 6.40 | 3.21 | 1.44 | 11.05 |
|   not exceeding 300mm | m | 0.10 | 1.60 | 1.09 | 0.40 | 3.09 |
| 32mm thick | | | | | | |
|   over 300mm wide | m2 | 0.44 | 7.04 | 3.67 | 1.61 | 12.32 |
|   not exceeding 300mm | m | 0.12 | 1.92 | 1.29 | 0.48 | 3.69 |
| 38mm thick | | | | | | |
|   over 300mm wide | m2 | 0.48 | 7.68 | 3.99 | 1.75 | 13.42 |
|   not exceeding 300mm | m | 0.14 | 2.24 | 1.94 | 0.63 | 4.81 |
| 50mm thick | | | | | | |
|   over 300mm wide | m2 | 0.52 | 8.32 | 4.89 | 1.98 | 15.19 |
|   not exceeding 300mm | m | 0.16 | 2.56 | 1.70 | 0.64 | 4.90 |
| 65mm thick | | | | | | |
|   over 300mm wide | m2 | 0.56 | 8.96 | 6.27 | 2.28 | 17.51 |
|   not exceeding 300mm | m | 0.18 | 2.88 | 2.20 | 0.76 | 5.84 |

Cement and sand (1:3) screed in beds with steel trowelled finish to level landings

| | Unit | Labour hours | Labour cost £ | Materials £ | O & P £ | Total £ |
|---|---|---|---|---|---|---|
| 18mm thick | | | | | | |
|   over 300mm wide | m2 | 0.30 | 4.80 | 2.71 | 1.13 | 8.64 |
|   not exceeding 300mm | m | 0.10 | 1.60 | 0.92 | 0.38 | 2.90 |

| | Unit | Labour hours | Labour cost £ | Materials £ | O & P £ | Total £ |
|---|---|---|---|---|---|---|
| 25mm thick | | | | | | |
| over 300mm wide | m2 | 0.40 | 6.40 | 3.21 | 1.44 | 11.05 |
| not exceeding 300mm | m | 0.10 | 1.60 | 1.05 | 0.40 | 3.05 |
| 32mm thick | | | | | | |
| over 300mm wide | m2 | 0.44 | 7.04 | 3.67 | 1.61 | 12.32 |
| not exceeding 300mm | m | 0.12 | 1.92 | 1.22 | 0.47 | 3.61 |
| 38mm thick | | | | | | |
| over 300mm wide | m2 | 0.48 | 7.68 | 3.99 | 1.75 | 13.42 |
| not exceeding 300mm | m | 0.14 | 2.24 | 1.32 | 0.53 | 4.09 |
| 50mm thick | | | | | | |
| over 300mm wide | m2 | 0.52 | 8.32 | 4.89 | 1.98 | 15.19 |
| not exceeding 300mm | m | 0.16 | 2.56 | 1.63 | 0.63 | 4.82 |
| 65mm thick | | | | | | |
| over 300mm wide | m2 | 0.56 | 8.96 | 6.27 | 2.28 | 17.51 |
| not exceeding 300mm | m | 0.18 | 2.88 | 2.37 | 0.79 | 6.04 |
| **Cement and sand (1:3) screed in beds with steel trowelled finish to risers and treads** | | | | | | |
| 25mm thick | | | | | | |
| 250mm wide | m | 0.34 | 5.44 | 0.81 | 0.94 | 7.19 |
| 350mm wide | m | 0.40 | 6.40 | 1.12 | 1.13 | 8.65 |
| 32mm thick | | | | | | |
| 250mm wide | m | 0.38 | 6.08 | 0.91 | 1.05 | 8.04 |
| 350mm wide | m | 0.44 | 7.04 | 1.10 | 1.22 | 9.36 |

| | Unit | Labour hours | Labour cost £ | Materials £ | O & P £ | Total £ |
|---|---|---|---|---|---|---|
| **Cement and sand (1:3) screed (cont'd)** | | | | | | |
| 38mm thick | | | | | | |
| 250mm wide | m | 0.42 | 6.72 | 1.00 | 1.16 | 8.88 |
| 350mm wide | m | 0.48 | 7.68 | 1.40 | 1.36 | 10.44 |
| 50mm thick | | | | | | |
| 250mm wide | m | 0.46 | 7.36 | 1.22 | 1.29 | 9.87 |
| 350mm wide | m | 0.52 | 8.32 | 1.72 | 1.51 | 11.55 |
| **Granolithic, cement and granite chippings (1:2:5) with steel trowelled finish to level floors or to falls not exceeding 15 degrees from the horizontal** | | | | | | |
| 25mm thick | | | | | | |
| over 300mm wide | m2 | 0.40 | 6.40 | 5.12 | 1.73 | 13.25 |
| not exceeding 300mm | m | 0.10 | 1.60 | 1.58 | 0.48 | 3.66 |
| 32mm thick | | | | | | |
| over 300mm wide | m2 | 0.44 | 7.04 | 6.46 | 2.03 | 15.53 |
| not exceeding 300mm | m | 0.12 | 1.92 | 1.69 | 0.54 | 4.15 |
| 38mm thick | | | | | | |
| over 300mm wide | m2 | 0.48 | 7.68 | 5.99 | 2.05 | 15.72 |
| not exceeding 300mm | m | 0.14 | 2.24 | 1.85 | 0.61 | 4.70 |
| 50mm thick | | | | | | |
| over 300mm wide | m2 | 0.52 | 8.32 | 6.93 | 2.29 | 17.54 |
| not exceeding 300mm | m | 0.16 | 2.56 | 2.14 | 0.71 | 5.41 |
| 65mm thick | | | | | | |
| over 300mm wide | m2 | 0.56 | 8.96 | 8.53 | 2.62 | 20.11 |
| not exceeding 300mm | m | 0.18 | 2.88 | 2.61 | 0.82 | 6.31 |

| | Unit | Labour hours | Labour cost £ | Materials £ | O & P £ | Total £ |
|---|---|---|---|---|---|---|
| **Granolithic, cement and granite chippings (1:2:5) with steel trowelled finish to level landings** | | | | | | |
| 25mm thick | | | | | | |
| over 300mm wide | m2 | 0.40 | 6.40 | 5.12 | 1.73 | 13.25 |
| not exceeding 300mm | m | 0.10 | 1.60 | 1.54 | 0.47 | 3.61 |
| 32mm thick | | | | | | |
| over 300mm wide | m2 | 0.44 | 7.04 | 5.55 | 1.89 | 14.48 |
| not exceeding 300mm | m | 0.12 | 1.92 | 1.66 | 0.54 | 4.12 |
| 38mm thick | | | | | | |
| over 300mm wide | m2 | 0.48 | 7.68 | 5.99 | 2.05 | 15.72 |
| not exceeding 300mm | m | 0.14 | 2.24 | 1.80 | 0.61 | 4.65 |
| 50mm thick | | | | | | |
| over 300mm wide | m2 | 0.52 | 8.32 | 6.95 | 2.29 | 17.56 |
| not exceeding 300mm | m | 0.16 | 2.56 | 2.09 | 0.70 | 5.35 |
| **Granolithic, cement and granite chippings (1:2:5) with steel trowelled finish to risers and treads** | | | | | | |
| 25mm thick | | | | | | |
| 250mm wide | m | 0.34 | 5.44 | 1.27 | 1.01 | 7.72 |
| 350mm wide | m | 0.40 | 6.40 | 1.79 | 1.23 | 9.42 |
| 32mm thick | | | | | | |
| 250mm wide | m2 | 0.38 | 6.08 | 1.38 | 1.12 | 8.58 |
| 350mm wide | m | 0.44 | 7.04 | 1.94 | 1.35 | 10.33 |
| 38mm thick | | | | | | |
| 250mm wide | m2 | 0.42 | 6.72 | 1.49 | 1.23 | 9.44 |
| 350mm wide | m | 0.48 | 7.68 | 2.10 | 1.47 | 11.25 |

| | Unit | Labour hours | Labour cost £ | Materials £ | O & P £ | Total £ |
|---|---|---|---|---|---|---|
| **In situ wall coverings** | | | | | | |
| Cement and sand (1:3) beds and backings to brickwork or block-work walls | | | | | | |
| 13mm thick | | | | | | |
| over 300mm wide | m2 | 0.50 | 8.00 | 1.46 | 1.42 | 10.88 |
| not exceeding 300mm | m | 0.20 | 3.20 | 0.52 | 0.56 | 4.28 |
| 19mm thick | | | | | | |
| over 300mm wide | m2 | 0.58 | 9.28 | 2.62 | 1.79 | 13.69 |
| not exceeding 300mm | m | 0.22 | 3.52 | 0.93 | 0.67 | 5.12 |
| Cement and sand (1:3) beds and backings to sides of isolated columns | | | | | | |
| 13mm thick | | | | | | |
| over 300mm wide | m2 | 0.60 | 9.60 | 1.46 | 1.66 | 12.72 |
| not exceeding 300mm | m | 0.24 | 3.84 | 0.52 | 0.65 | 5.01 |
| 19mm thick | | | | | | |
| over 300mm wide | m2 | 0.68 | 10.88 | 2.62 | 2.03 | 15.53 |
| not exceeding 300mm | m | 0.26 | 4.16 | 0.93 | 0.76 | 5.85 |
| Premixed lightweight plaster, 11mm bonding, 2mm finish | | | | | | |
| brickwork or blockwork walls | | | | | | |
| over 300mm wide | m2 | 0.48 | 7.68 | 2.87 | 1.58 | 12.13 |
| not exceeding 300mm | m | 0.18 | 2.88 | 0.97 | 0.58 | 4.43 |
| isolated columns | | | | | | |
| over 300mm wide | m2 | 0.58 | 9.28 | 2.87 | 1.82 | 13.97 |
| not exceeding 300mm | m | 0.22 | 3.52 | 0.97 | 0.67 | 5.16 |
| ceilings | | | | | | |
| over 300mm wide | m2 | 0.62 | 9.92 | 2.87 | 1.92 | 14.71 |
| not exceeding 300mm | m | 0.24 | 3.84 | 0.97 | 0.72 | 5.53 |

| | Unit | Labour hours | Labour cost £ | Materials £ | O & P £ | Total £ |
|---|---|---|---|---|---|---|
| **One coat 'Universal' plaster to brickwork or blockwork walls** | | | | | | |
| 13mm thick | | | | | | |
|   over 300mm wide | m2 | 0.40 | 6.40 | 3.55 | 1.49 | 11.44 |
|   not exceeding 300mm | m | 0.15 | 2.40 | 1.17 | 0.54 | 4.11 |
| 19mm thick | | | | | | |
|   over 300mm wide | m2 | 0.45 | 7.20 | 4.52 | 1.76 | 13.48 |
|   not exceeding 300mm | m | 0.18 | 2.88 | 2.00 | 0.73 | 5.61 |
| **One coat 'Universal' plaster to isolated columns** | | | | | | |
| 13mm thick | | | | | | |
|   over 300mm wide | m2 | 0.44 | 7.04 | 3.65 | 1.60 | 12.29 |
|   not exceeding 300mm | m | 0.18 | 2.88 | 1.17 | 0.61 | 4.66 |
| 19mm thick | | | | | | |
|   over 300mm wide | m2 | 0.48 | 7.68 | 4.52 | 1.83 | 14.03 |
|   not exceeding 300mm | m | 0.20 | 3.20 | 1.23 | 0.66 | 5.09 |
| **One coat board finish 3mm thick** | | | | | | |
| plasterboard walls | | | | | | |
|   over 300mm wide | m2 | 0.30 | 4.80 | 1.73 | 0.98 | 7.51 |
|   not exceeding 300mm | m | 0.12 | 1.92 | 0.63 | 0.38 | 2.93 |
| plasterboard ceilings | | | | | | |
|   over 300mm wide | m2 | 0.40 | 6.40 | 1.73 | 1.22 | 9.35 |
|   not exceeding 300mm | m | 0.14 | 2.24 | 0.63 | 0.43 | 3.30 |
| plasterboard isolated columns | | | | | | |
|   over 300mm wide | m2 | 0.44 | 7.04 | 1.73 | 1.32 | 10.09 |
|   not exceeding 300mm | m | 0.18 | 2.88 | 0.63 | 0.53 | 4.04 |

|  | Unit | Labour hours | Labour cost £ | Materials £ | O & P £ | Total £ |
|---|---|---|---|---|---|---|
| **Lathing** | | | | | | |
| Expamet expanded metal lathing, 6mm thick | | | | | | |
| stapled to softwood, vertically | | | | | | |
| over 300mm wide | m2 | 0.26 | 4.16 | 4.65 | 1.32 | 10.13 |
| not exceeding 300mm | m | 0.09 | 1.44 | 1.47 | 0.44 | 3.35 |
| tied with wire, vertically | | | | | | |
| over 300mm wide | m2 | 0.29 | 4.64 | 4.65 | 1.39 | 10.68 |
| not exceeding 300mm | m | 0.11 | 1.76 | 1.47 | 0.48 | 3.71 |
| tied with wire, horizontally | | | | | | |
| over 300mm wide | m2 | 0.32 | 5.12 | 4.65 | 1.47 | 11.24 |
| not exceeding 300mm | m | 0.12 | 1.92 | 1.47 | 0.51 | 3.90 |
| Expamet Riblath expanded metal lathing, 0.3mm thick | | | | | | |
| stapled to softwood, vertically | | | | | | |
| over 300mm wide | m2 | 0.28 | 4.48 | 5.79 | 1.54 | 11.81 |
| not exceeding 300mm | m | 0.10 | 1.60 | 1.96 | 0.53 | 4.09 |
| tied with wire, vertically | | | | | | |
| over 300mm wide | m2 | 0.31 | 4.96 | 5.79 | 1.61 | 12.36 |
| not exceeding 300mm | m | 0.12 | 1.92 | 1.96 | 0.58 | 4.46 |
| tied with wire, horizontally | | | | | | |
| over 300mm wide | m2 | 0.34 | 5.44 | 5.79 | 1.68 | 12.91 |
| not exceeding 300mm | m | 0.13 | 2.08 | 1.96 | 0.61 | 4.65 |

| | Unit | Labour hours | Labour cost £ | Materials £ | O & P £ | Total £ |
|---|---|---|---|---|---|---|

**Plasterboard**

Tapered edge wallboard, open
joints to receive skimming or
similar, over 300mm wide

9.5mm thick to

| | Unit | Labour hours | Labour cost £ | Materials £ | O & P £ | Total £ |
|---|---|---|---|---|---|---|
| walls | m2 | 0.28 | 4.48 | 2.50 | 1.05 | 8.03 |
| ceilings | m2 | 0.30 | 4.80 | 2.50 | 1.10 | 8.40 |
| beams | m2 | 0.33 | 5.28 | 2.50 | 1.17 | 8.95 |
| columns | m2 | 0.33 | 5.28 | 2.50 | 1.17 | 8.95 |

12.5mm thick to

| | Unit | Labour hours | Labour cost £ | Materials £ | O & P £ | Total £ |
|---|---|---|---|---|---|---|
| walls | m2 | 0.30 | 4.80 | 2.98 | 1.17 | 8.95 |
| ceilings | m2 | 0.35 | 5.60 | 2.98 | 1.29 | 9.87 |
| beams | m2 | 0.40 | 6.40 | 2.98 | 1.41 | 10.79 |
| columns | m2 | 0.40 | 6.40 | 2.98 | 1.41 | 10.79 |

Tapered edge wallboard, open
joints to receive skimming or
similar, not exceeding
300mm wide

9.5mm thick to

| | Unit | Labour hours | Labour cost £ | Materials £ | O & P £ | Total £ |
|---|---|---|---|---|---|---|
| walls | m | 0.16 | 2.56 | 2.34 | 1.91 | 6.81 |
| ceilings | m | 0.35 | 5.60 | 2.34 | 1.91 | 9.85 |
| beams | m | 0.35 | 5.60 | 2.34 | 1.91 | 9.85 |
| columns | m | 0.40 | 6.40 | 2.82 | 1.91 | 11.13 |

12.5mm thick to

| | Unit | Labour hours | Labour cost £ | Materials £ | O & P £ | Total £ |
|---|---|---|---|---|---|---|
| walls | m | 0.35 | 5.60 | 2.34 | 2.02 | 9.96 |
| ceilings | m | 0.35 | 5.60 | 2.34 | 2.02 | 9.96 |
| beams | m | 0.35 | 5.60 | 2.34 | 2.02 | 9.96 |
| columns | m | 0.40 | 6.40 | 2.82 | 2.02 | 11.24 |

| | Unit | Labour hours | Labour cost £ | Materials £ | O & P £ | Total £ |
|---|---|---|---|---|---|---|
| **Floor tiling** | | | | | | |
| Red clay quarry tiles, bedded in cement mortar (1:3) 12mm thick, butt jointed straight both ways | | | | | | |
| 150 × 150 × 12.5mm thick to floors | | | | | | |
| over 300mm wide | m2 | 0.90 | 14.40 | 21.42 | 5.37 | 41.19 |
| not exceeding 300mm | m | 0.36 | 5.76 | 8.03 | 2.07 | 15.86 |
| 225 × 225 × 25mm thick to floors | | | | | | |
| over 300mm wide | m2 | 0.80 | 12.80 | 29.92 | 6.41 | 49.13 |
| not exceeding 300mm | m | 0.32 | 5.12 | 9.71 | 2.22 | 17.05 |
| Vitrified ceramic tiles, bedded in cement mortar (1:3) 12mm thick, butt jointed straight both ways | | | | | | |
| 100 × 100 × 9mm thick to floors | | | | | | |
| over 300mm wide | m2 | 1.00 | 16.00 | 20.50 | 5.48 | 41.98 |
| not exceeding 300mm | m | 0.44 | 7.04 | 7.96 | 2.25 | 17.25 |
| 150 × 150 × 12.5mm thick to floors | | | | | | |
| over 300mm wide | m2 | 0.90 | 14.40 | 22.67 | 5.56 | 42.63 |
| not exceeding 300mm | m | 0.36 | 5.76 | 8.19 | 2.09 | 16.04 |
| 200 × 200 × 12.5mm thick to floors | | | | | | |
| over 300mm wide | m2 | 0.80 | 12.80 | 23.76 | 5.48 | 42.04 |
| not exceeding 300mm | m | 0.32 | 5.12 | 8.81 | 2.09 | 16.02 |

|  | Unit | Labour hours | Labour cost £ | Materials £ | O & P £ | Total £ |
|---|---|---|---|---|---|---|
| **Vinyl floor tiling size 300 × 300mm fixed with adhesive** | | | | | | |
| 2mm thick | | | | | | |
| over 300mm wide | m2 | 0.28 | 4.48 | 12.81 | 2.59 | 19.88 |
| not exceeding 300mm | m | 0.10 | 1.60 | 4.95 | 0.98 | 7.53 |
| 2.5mm thick | | | | | | |
| over 300mm wide | m2 | 0.28 | 4.48 | 13.82 | 2.75 | 21.05 |
| not exceeding 300mm | m | 0.10 | 1.60 | 5.06 | 1.00 | 7.66 |
| 3mm thick | | | | | | |
| over 300mm wide | m2 | 0.32 | 5.12 | 14.53 | 2.95 | 22.60 |
| not exceeding 300mm | m | 0.12 | 1.92 | 5.84 | 1.16 | 8.92 |
| **Vinyl floor sheeting with welded joints, fixed with adhesive** | | | | | | |
| 2mm thick | | | | | | |
| over 300mm wide | m2 | 0.36 | 5.76 | 12.69 | 2.77 | 21.22 |
| not exceeding 300mm | m | 0.14 | 2.24 | 4.88 | 1.07 | 8.19 |
| 2.5mm thick | | | | | | |
| over 300mm wide | m2 | 0.40 | 6.40 | 13.04 | 2.92 | 22.36 |
| not exceeding 300mm | m | 0.18 | 2.88 | 5.01 | 1.18 | 9.07 |
| **Linoleum floor sheeting with butt joints, fixed to adhesive** | | | | | | |
| 2mm thick | | | | | | |
| over 300mm wide | m2 | 0.36 | 5.76 | 14.79 | 3.08 | 23.63 |
| not exceeding 300mm | m | 0.14 | 2.24 | 6.59 | 1.32 | 10.15 |
| 2.5mm thick | | | | | | |
| over 300mm wide | m2 | 0.36 | 5.76 | 18.40 | 3.62 | 27.78 |
| not exceeding 300mm | m | 0.14 | 2.24 | 7.81 | 1.51 | 11.56 |

| | Unit | Labour hours | Labour cost £ | Materials £ | O & P £ | Total £ |
|---|---|---|---|---|---|---|
| **Wall tiling** | | | | | | |
| White glazed ceramic wall tiling, fixed with adhesive, pointing with white grout | | | | | | |
| 108 × 108 × 5.5mm thick | | | | | | |
| over 300mm wide | m2 | 0.95 | 15.20 | 16.50 | 4.76 | 36.46 |
| not exceeding 300mm | m | 0.38 | 6.08 | 4.12 | 1.53 | 11.73 |
| 108 × 108 × 6.5mm thick | | | | | | |
| over 300mm wide | m2 | 0.80 | 12.80 | 17.04 | 4.48 | 34.32 |
| not exceeding 300mm | m | 0.35 | 5.60 | 4.32 | 1.49 | 11.41 |
| 152 × 152 × 6.5mm thick | | | | | | |
| over 300mm wide | m2 | 0.75 | 12.00 | 16.21 | 4.23 | 32.44 |
| not exceeding 300mm | m | 0.32 | 5.12 | 4.22 | 1.40 | 10.74 |
| Extra for | | | | | | |
| cutting to edges | m | 0.10 | 1.60 | 0.00 | 0.24 | 1.84 |
| holes for small pipes | nr | 0.20 | 3.20 | 0.00 | 0.48 | 3.68 |
| holes for large pipes | nr | 0.25 | 4.00 | 0.00 | 0.60 | 4.60 |
| Dark-coloured glazed ceramic wall tiling, fixed with adhesive, pointing with matching grout | | | | | | |
| 108 × 108 × 6.5mm thick | | | | | | |
| over 300mm wide | m2 | 0.95 | 15.20 | 19.03 | 5.13 | 39.36 |
| not exceeding 300mm | m | 0.38 | 6.08 | 5.02 | 1.67 | 12.77 |

| | Unit | Labour hours | Labour cost £ | Materials £ | O & P £ | Total £ |
|---|---|---|---|---|---|---|
| **152 × 152 × 5.5mm thick** | | | | | | |
| over 300mm wide | m2 | 0.80 | 12.80 | 18.61 | 4.71 | 36.12 |
| not exceeding 300mm | m | 0.35 | 5.60 | 4.87 | 1.57 | 12.04 |
| **203 × 102 × 6.5mm thick** | | | | | | |
| over 300mm wide | m2 | 0.75 | 12.00 | 19.50 | 4.73 | 36.23 |
| not exceeding 300mm | m | 0.32 | 5.12 | 7.74 | 1.93 | 14.79 |
| Extra for | | | | | | |
| cutting to edges | m | 0.10 | 1.60 | 0.00 | 0.24 | 1.84 |
| holes for small pipes | nr | 0.20 | 3.20 | 0.00 | 0.48 | 3.68 |
| holes for large pipes | nr | 0.25 | 4.00 | 0.00 | 0.60 | 4.60 |

| | Unit | Labour hours | Labour cost £ | Materials £ | O & P £ | Total £ |
|---|---|---|---|---|---|---|

## PLUMBING AND HEATING

### Sanitary fittings

Acrylic reinforced bath size 1500
× 700mm complete with 2nr
chromium-plated grips, 40mm
waste fitting, overflow with chain
and plastic plug

| | Unit | Labour hours | Labour cost £ | Materials £ | O & P £ | Total £ |
|---|---|---|---|---|---|---|
| white | nr | 2.75 | 48.13 | 135.77 | 27.58 | 211.48 |

Acrylic reinforced bath size 1700
× 700mm complete with 2nr
chromium-plated grips, 40mm
waste fitting, overflow with chain
and plastic plug

| | Unit | Labour hours | Labour cost £ | Materials £ | O & P £ | Total £ |
|---|---|---|---|---|---|---|
| white | nr | 2.90 | 50.75 | 153.94 | 30.70 | 235.39 |
| coloured | nr | 2.90 | 50.75 | 167.65 | 32.76 | 251.16 |

Porcelain enamel standard gauge
bath size 1700 × 700mm complete
with 2nr chromium-plated grips,
40mm waste fitting, overflow
with chain and plastic plug

| | Unit | Labour hours | Labour cost £ | Materials £ | O & P £ | Total £ |
|---|---|---|---|---|---|---|
| white | nr | 3.00 | 52.50 | 161.25 | 32.06 | 245.81 |
| coloured | nr | 3.00 | 52.50 | 175.79 | 34.24 | 262.53 |

Porcelain enamel heavy gauge
bath size 1700 × 700mm complete
with 2nr chromium-plated grips,
40mm waste fitting, overflow
with chain and plastic plug

| | Unit | Labour hours | Labour cost £ | Materials £ | O & P £ | Total £ |
|---|---|---|---|---|---|---|
| white | nr | 3.10 | 54.25 | 168.59 | 33.43 | 256.27 |
| coloured | nr | 3.10 | 54.25 | 212.21 | 39.97 | 306.43 |

| | Unit | Labour hours | Labour cost £ | Materials £ | O & P £ | Total £ |
|---|---|---|---|---|---|---|
| Porcelain enamel heavy gauge shallow bath size 1700 × 700mm complete with 2nr chromium-plated grips, 40mm waste fitting, overflow with chain and plastic plug | | | | | | |
| white | nr | 3.10 | 54.25 | 248.67 | 45.44 | 348.36 |
| coloured | nr | 3.10 | 54.25 | 289.43 | 51.55 | 395.23 |
| Bath panels, moulded acrylic, fixing in position for trimming as required | | | | | | |
| end panel | nr | 0.30 | 5.25 | 13.67 | 2.84 | 21.76 |
| side panel | nr | 0.50 | 8.75 | 16.33 | 3.76 | 28.84 |
| **Bath accessories** | | | | | | |
| Angle strip polished aluminium, fixing with chromium-plated dome headed screws, cutting to length | | | | | | |
| 25 × 25 × 560mm long | nr | 0.35 | 6.13 | 4.95 | 1.66 | 12.74 |
| Wash basin, vitreous china size 510 × 410mm complete with pedestal, chromium-plated waste, overflow with chain and plastic plug | | | | | | |
| white | nr | 2.20 | 38.50 | 89.88 | 19.26 | 147.64 |
| coloured | nr | 2.20 | 38.50 | 97.82 | 20.45 | 156.77 |
| **Basins and pedestals** | | | | | | |
| Wash basin, vitreous china size 560 × 455mm complete with pedestal, chromium-plated waste, overflow with chain and plastic plug | | | | | | |
| white | nr | 2.25 | 39.38 | 82.45 | 18.27 | 140.10 |
| coloured | nr | 2.25 | 39.38 | 99.78 | 20.87 | 160.03 |

| | Unit | Labour hours | Labour cost £ | Materials £ | O & P £ | Total £ |
|---|---|---|---|---|---|---|
| Wash basin, vitreous china size 610 × 470mm complete with pedestal, chromium-plated waste, overflow with chain and plastic plug | | | | | | |
| white | nr | 2.30 | 40.25 | 93.31 | 20.03 | 153.59 |
| coloured | nr | 2.30 | 40.25 | 100.87 | 21.17 | 162.29 |
| Wash basin, vitreous china size 510 × 410mm complete with brackets, chromium-plated waste, overflow with chain and plastic plug | | | | | | |
| white | nr | 2.00 | 35.00 | 69.22 | 15.63 | 119.85 |
| coloured | nr | 2.00 | 35.00 | 79.48 | 17.17 | 131.65 |
| Wash basin, vitreous china size 560 × 455mm complete with brackets, chromium-plated waste, overflow with chain and plastic plug | | | | | | |
| white | nr | 2.25 | 39.38 | 71.20 | 16.59 | 127.16 |
| coloured | nr | 2.25 | 39.38 | 82.41 | 18.27 | 140.05 |
| Sinks and tops | | | | | | |
| Stainless steel single sit-on type sink, complete with inset chromium-plated waste, overflow with chain and plastic plug, size | | | | | | |
| 1000 × 500mm | nr | 1.30 | 22.75 | 94.00 | 17.51 | 134.26 |
| 1000 × 600mm | nr | 1.40 | 24.50 | 100.98 | 18.82 | 144.30 |
| Stainless steel single roll-edge type sink, complete with inset chromium-plated waste, overflow with chain and plastic plug, size | | | | | | |
| 1000 × 500mm | nr | 1.30 | 22.75 | 97.06 | 17.97 | 137.78 |
| 1000 × 600mm | nr | 1.40 | 24.50 | 105.30 | 19.47 | 149.27 |
| 1200 × 600mm | nr | 1.50 | 26.25 | 117.76 | 21.60 | 165.61 |

| | Unit | Labour hours | Labour cost £ | Materials £ | O & P £ | Total £ |
|---|---|---|---|---|---|---|
| Stainless steel double sit-on type sink, complete with inset chromium-plated waste, overflow with chain and plastic plug, size | | | | | | |
| 1500 × 500mm | nr | 1.80 | 31.50 | 117.21 | 22.31 | 171.02 |
| 1500 × 600mm | nr | 1.90 | 33.25 | 134.76 | 25.20 | 193.21 |
| Stainless steel double roll-edge type sink, complete with inset chromium-plated waste, overflow with chain and plastic plug, size | | | | | | |
| 1000 × 500mm | nr | 1.80 | 31.50 | 126.12 | 23.64 | 181.26 |
| 1000 × 600mm | nr | 1.90 | 33.25 | 140.18 | 26.01 | 199.44 |
| Belfast pattern white fireclay sink, complete with chromium-plated waste, chain and plastic plug, wall-mounted on brackets, size | | | | | | |
| 450 × 380 × 205mm | nr | 2.35 | 41.13 | 125.97 | 25.06 | 192.16 |
| 610 × 455 × 255mm | nr | 2.40 | 42.00 | 175.74 | 32.66 | 250.40 |
| 760 × 455 × 255mm | nr | 2.45 | 42.88 | 236.05 | 41.84 | 320.76 |

WC suites

| | Unit | Labour hours | Labour cost £ | Materials £ | O & P £ | Total £ |
|---|---|---|---|---|---|---|
| Vitreous white china low-level WC suite comprising pan, plastic seat and cover, 9 litre cistern, low-pressure ball valve, connecting pipework screwed to floor | nr | 2.35 | 41.13 | 222.97 | 39.61 | 303.71 |

| | Unit | Labour hours | Labour cost £ | Materials £ | O & P £ | Total £ |
|---|---|---|---|---|---|---|
| Vitreous coloured china low-level WC suite comprising pan, plastic seat and cover, 9 litre cistern, low-pressure ball valve, connecting pipework screwed to floor | nr | 2.35 | 41.13 | 253.78 | 44.24 | 339.14 |
| Vitreous white china low-level close coupled WC suite comprising pan, plastic seat and cover, 9 litre cistern, low-pressure ball valve, connecting pipework screwed to floor | nr | 2.45 | 42.88 | 299.71 | 51.39 | 393.97 |
| **Bidets** | | | | | | |
| Free-standing plain rim, vitreous china bidet excluding fittings | | | | | | |
| white | nr | 2.50 | 43.75 | 282.08 | 48.87 | 374.70 |
| coloured | nr | 2.50 | 43.75 | 303.45 | 52.08 | 399.28 |
| **Urinals** | | | | | | |
| Bowl urinals, white vitreous china, wall-mounted on hangers chromium-plated dome outlet 430mm wide × 305mm high | nr | 1.00 | 17.50 | 120.05 | 20.63 | 158.18 |
| Stainless steel flush pipes and spreaders for bowl urinals, face fixed to wall | | | | | | |
| one bowl set | nr | 1.20 | 21.00 | 64.27 | 12.79 | 98.06 |
| two bowl set | nr | 1.75 | 30.63 | 111.68 | 21.35 | 163.65 |
| three bowl set | nr | 2.10 | 36.75 | 162.32 | 29.86 | 228.93 |
| Automatic flushing cistern and fittings, white vitreous china, wall-mounted on brackets | | | | | | |
| 4.5 litre | nr | 1.00 | 17.50 | 135.98 | 23.02 | 176.50 |
| 9 litre | nr | 1.05 | 18.38 | 152.89 | 25.69 | 196.95 |
| 13.5 litre | nr | 1.15 | 20.13 | 184.94 | 30.76 | 235.82 |

|  | Unit | Labour hours | Labour cost £ | Materials £ | O & P £ | Total £ |
|---|---|---|---|---|---|---|
| **Automatic flushing cistern and fittings, coloured vitreous china, wall-mounted on brackets** | | | | | | |
| 4.5 litre | nr | 1.00 | 17.50 | 151.37 | 25.33 | 194.20 |
| 9 litre | nr | 1.05 | 18.38 | 176.55 | 29.24 | 224.16 |
| 13.5 litre | nr | 1.15 | 20.13 | 199.67 | 32.97 | 252.76 |
| **Modular slab urinal, white fireclay china, comprising back slabs without divisions, waterway channel, automatic flushing cistern, stainless steel flush pipes** | | | | | | |
| two persons | nr | 5.20 | 91.00 | 656.47 | 112.12 | 859.59 |
| three persons | nr | 6.00 | 105.00 | 1076.57 | 177.24 | 1358.81 |
| four persons | nr | 7.00 | 122.50 | 1363.23 | 222.86 | 1708.59 |
| **Taps** | | | | | | |
| **Chromium-plated pillar taps** | | | | | | |
| 13mm | pr | 0.40 | 7.00 | 26.02 | 4.95 | 37.97 |
| 19mm | pr | 0.45 | 7.88 | 34.79 | 6.40 | 49.06 |
| **Chromium-plated mixer taps, 19mm** | | | | | | |
| cross-top handles | pr | 0.60 | 10.50 | 64.23 | 11.21 | 85.94 |
| lever handles | pr | 0.60 | 10.50 | 86.52 | 14.55 | 111.57 |
| **Chromium-plated bidet monoblock and spray** | nr | 0.70 | 12.25 | 185.58 | 29.67 | 227.50 |

| | Unit | Labour hours | Labour cost £ | Materials £ | O & P £ | Total £ |
|---|---|---|---|---|---|---|
| **Valves** | | | | | | |
| Chromium-plated thermostatic exposed shower valve with flexible hose, slide rail and spray set | nr | 1.40 | 24.50 | 219.70 | 36.63 | 280.83 |
| Instant electric 9.5kw shower with flexible hose, slide rail and spray set | nr | 1.50 | 26.25 | 182.33 | 31.29 | 239.87 |
| **Showers** | | | | | | |
| Shower cubicle size 788 × 842 × 2115mm in anodised aluminium frame and safety glass | nr | 2.00 | 35.00 | 751.53 | 117.98 | 904.51 |
| White glazed fireclay shower tray size 900 × 900 × 180mm with chromium-plated waste fitting | nr | 1.80 | 31.50 | 194.79 | 33.94 | 260.23 |
| White acrylic fireclay shower tray size 750 × 750 × 180mm with chromium-plated waste fitting | nr | 1.30 | 22.75 | 102.84 | 18.84 | 144.43 |
| **Holes and chases** | | | | | | |
| Cutting holes for pipes up to 25 to 50mm diameter in walls 102.5mm thick | | | | | | |
| commons | nr | 0.50 | 8.75 | 0.00 | 1.31 | 10.06 |
| facings | nr | 0.80 | 14.00 | 0.00 | 2.10 | 16.10 |
| engineering class A | nr | 1.00 | 17.50 | 0.00 | 2.63 | 20.13 |
| engineering class B | nr | 0.85 | 14.88 | 0.00 | 2.23 | 17.11 |

| | Unit | Labour hours | Labour cost £ | Materials £ | O & P £ | Total £ |
|---|---|---|---|---|---|---|
| **Cutting holes for pipes up to 25 to 50mm diameter in walls 100mm thick** | | | | | | |
| concrete blocks | nr | 0.40 | 7.00 | 0.00 | 1.05 | 8.05 |
| Thermalite blocks | nr | 0.30 | 5.25 | 0.00 | 0.79 | 6.04 |
| aerated concrete blocks | nr | 0.35 | 6.13 | 0.00 | 0.92 | 7.04 |
| **Cutting holes for pipes up to 25 to 50mm diameter in walls 225mm thick** | | | | | | |
| commons | nr | 1.05 | 18.38 | 0.00 | 2.76 | 21.13 |
| facings | nr | 1.10 | 19.25 | 0.00 | 2.89 | 22.14 |
| engineering class A | nr | 1.60 | 28.00 | 0.00 | 4.20 | 32.20 |
| engineering class B | nr | 1.80 | 31.50 | 0.00 | 4.73 | 36.23 |
| **Cutting holes for pipes up to 25 to 50mm diameter in walls 140mm thick** | | | | | | |
| concrete blocks | nr | 0.60 | 10.50 | 0.00 | 1.58 | 12.08 |
| Thermalite blocks | nr | 0.40 | 7.00 | 0.00 | 1.05 | 8.05 |
| aerated concrete blocks | nr | 0.45 | 7.88 | 0.00 | 1.18 | 9.06 |
| **Cutting holes for pipes up to 50 to 75mm diameter in walls 102.5mm thick** | | | | | | |
| commons | nr | 0.80 | 8.75 | 0.00 | 1.31 | 10.06 |
| facings | nr | 1.10 | 14.00 | 0.00 | 2.10 | 16.10 |
| engineering class A | nr | 1.15 | 17.50 | 0.00 | 2.63 | 20.13 |
| engineering class B | nr | 0.85 | 14.88 | 0.00 | 2.23 | 17.11 |
| **Cutting holes for pipes up to 50 to 75mm diameter in walls 100mm thick** | | | | | | |
| concrete blocks | nr | 0.70 | 7.00 | 0.00 | 1.05 | 8.05 |
| Thermalite blocks | nr | 0.60 | 5.25 | 0.00 | 0.79 | 6.04 |
| aerated concrete blocks | nr | 0.65 | 6.13 | 0.00 | 0.92 | 7.04 |

| | Unit | Labour hours | Labour cost £ | Materials £ | O & P £ | Total £ |
|---|---|---|---|---|---|---|
| Cutting holes for pipes up to 50 to 75mm diameter in walls 225mm thick | | | | | | |
| commons | nr | 1.35 | 18.38 | 0.00 | 2.76 | 21.13 |
| facings | nr | 1.40 | 19.25 | 0.00 | 2.89 | 22.14 |
| engineering class A | nr | 1.90 | 28.00 | 0.00 | 4.20 | 32.20 |
| engineering class B | nr | 2.10 | 31.50 | 0.00 | 4.73 | 36.23 |
| Cutting holes for pipes up to 50 to 75mm diameter in walls 140mm thick | | | | | | |
| concrete blocks | nr | 0.90 | 10.50 | 0.00 | 1.58 | 12.08 |
| Thermalite blocks | nr | 0.70 | 7.00 | 0.00 | 1.05 | 8.05 |
| aerated concrete blocks | nr | 0.75 | 7.88 | 0.00 | 1.18 | 9.06 |
| Cutting and pinning ends of pipe support brackets and make good to | | | | | | |
| concrete | nr | 0.45 | 7.88 | 0.00 | 1.18 | 9.06 |
| brickwork | nr | 0.35 | 6.13 | 0.00 | 0.92 | 7.04 |
| blockwork | nr | 0.30 | 5.25 | 0.00 | 0.79 | 6.04 |
| tiled walls | nr | 0.50 | 8.75 | 0.00 | 1.31 | 10.06 |
| Cut opening through cavity wall comprising facing bricks and concrete blocks | | | | | | |
| balanced flue outlet | nr | 0.90 | 15.75 | 0.00 | 2.36 | 18.11 |
| 150mm diameter flue pipe | nr | 0.75 | 13.13 | 0.00 | 1.97 | 15.09 |
| 19mm overflow pipe | nr | 0.20 | 3.50 | 0.00 | 0.53 | 4.03 |
| Cutting chases for one pipe up to 25mm diameter in | | | | | | |
| commons | nr | 0.35 | 6.13 | 0.00 | 0.92 | 7.04 |
| facings | nr | 0.38 | 6.65 | 0.00 | 1.00 | 7.65 |
| engineering class A | nr | 0.45 | 7.88 | 0.00 | 1.18 | 9.06 |
| engineering class B | nr | 0.40 | 7.00 | 0.00 | 1.05 | 8.05 |

|  | Unit | Labour hours | Labour cost £ | Materials £ | O & P £ | Total £ |
|---|---|---|---|---|---|---|
| concrete blocks | nr | 0.35 | 6.13 | 0.00 | 0.92 | 7.04 |
| Thermalite blocks | nr | 0.25 | 4.38 | 0.00 | 0.66 | 5.03 |
| aerated concrete blocks | nr | 0.20 | 3.50 | 0.00 | 0.53 | 4.03 |
| Cutting chases for two pipes up to 25mm diameter or one pipe 50mm diameter in |  |  |  |  |  |  |
| commons | m | 0.45 | 7.88 | 0.00 | 1.18 | 9.06 |
| facings | m | 0.50 | 8.75 | 0.00 | 1.31 | 10.06 |
| engineering class A | m | 0.65 | 11.38 | 0.00 | 1.71 | 13.08 |
| engineering class B | m | 0.55 | 9.63 | 0.00 | 1.44 | 11.07 |
| concrete blocks | m | 0.40 | 7.00 | 0.00 | 1.05 | 8.05 |
| Thermalite blocks | m | 0.30 | 5.25 | 0.00 | 0.79 | 6.04 |
| aerated concrete blocks | m | 0.90 | 15.75 | 0.00 | 2.36 | 18.11 |
| Make good surfaces one side of chases |  |  |  |  |  |  |
| plastered surfaces | m | 0.15 | 2.63 | 0.70 | 0.50 | 3.82 |
| tiled surfaces | m | 0.25 | 4.38 | 2.33 | 1.01 | 7.71 |
| concrete floor | m | 0.35 | 6.13 | 0.70 | 1.02 | 7.85 |
| granolithic floor | m | 0.50 | 8.75 | 0.75 | 1.43 | 10.93 |
| concrete soffit | m | 0.55 | 9.63 | 0.70 | 1.55 | 11.87 |
| Make good surfaces both sides of chases |  |  |  |  |  |  |
| plastered surfaces | m | 0.10 | 1.75 | 0.70 | 0.50 | 3.82 |
| tiled surfaces | m | 0.15 | 2.63 | 2.33 | 0.50 | 3.82 |
| concrete floor | m | 0.25 | 4.38 | 0.70 | 0.50 | 3.82 |
| granolithic floor | m | 0.35 | 6.13 | 0.75 | 0.50 | 3.82 |
| concrete soffit | m | 0.45 | 7.88 | 0.70 | 0.50 | 3.82 |

| | Unit | Labour hours | Labour cost £ | Materials £ | O & P £ | Total £ |
|---|---|---|---|---|---|---|

**Rainwater goods**

PVC-U rainwater pipe plugged
to brickwork with pipe brackets
and fitting clips at 2m maximum
centres

| | Unit | Labour hours | Labour cost £ | Materials £ | O & P £ | Total £ |
|---|---|---|---|---|---|---|
| 68mm diameter pipe | m | 0.25 | 4.38 | 6.08 | 1.57 | 12.02 |
| extra over for | | | | | | |
| bend, 87.5 degrees | nr | 0.25 | 4.38 | 5.84 | 1.53 | 11.75 |
| offset | nr | 0.25 | 4.38 | 3.48 | 1.18 | 9.03 |
| branch | nr | 0.25 | 4.38 | 11.64 | 2.40 | 18.42 |
| shoe | nr | 0.25 | 4.38 | 5.04 | 1.41 | 10.83 |
| access pipe | nr | 0.25 | 4.38 | 16.02 | 3.06 | 23.45 |
| hopper head | nr | 0.25 | 4.38 | 18.72 | 3.46 | 26.56 |
| 68mm square pipe | m | 0.25 | 4.38 | 7.29 | 1.75 | 13.41 |
| extra over for | | | | | | |
| bend, 87.5 degrees | nr | 0.25 | 4.38 | 3.73 | 1.22 | 9.32 |
| offset | nr | 0.25 | 4.38 | 6.55 | 1.64 | 12.56 |
| branch | nr | 0.25 | 4.38 | 13.19 | 2.63 | 20.20 |
| shoe | nr | 0.25 | 4.38 | 4.28 | 1.30 | 9.95 |
| access pipe | nr | 0.25 | 4.38 | 20.00 | 3.66 | 28.03 |
| hopper head | nr | 0.25 | 4.38 | 18.72 | 3.46 | 26.56 |

Aluminium rainwater pipe,
straight, plain eared, spigot and
socket dry joints plugged to
brickwork

| | Unit | Labour hours | Labour cost £ | Materials £ | O & P £ | Total £ |
|---|---|---|---|---|---|---|
| 63mm diameter pipe | m | 0.28 | 4.90 | 16.63 | 3.23 | 24.76 |
| extra over for | | | | | | |
| offset, 75mm | nr | 0.28 | 4.90 | 31.47 | 5.46 | 41.83 |
| offset, 100mm | nr | 0.28 | 4.90 | 31.47 | 5.46 | 41.83 |
| offset, 150mm | nr | 0.28 | 4.90 | 31.47 | 5.46 | 41.83 |
| offset, 225mm | nr | 0.28 | 4.90 | 37.77 | 6.40 | 49.07 |
| offset, 300mm | nr | 0.28 | 4.90 | 37.77 | 6.40 | 49.07 |

| | Unit | Labour hours | Labour cost £ | Materials £ | O & P £ | Total £ |
|---|---|---|---|---|---|---|
| bend, 92.5 degrees | nr | 0.28 | 4.90 | 25.61 | 4.58 | 35.09 |
| branch | nr | 0.28 | 4.90 | 27.40 | 4.85 | 37.15 |
| shoe | nr | 0.28 | 4.90 | 9.54 | 2.17 | 16.61 |
| standard hopper | nr | 0.28 | 4.90 | 26.61 | 4.73 | 36.24 |
| flatback hopper | nr | 0.28 | 4.90 | 18.59 | 3.52 | 27.01 |
| 76mm diameter pipe | m | 0.30 | 5.25 | 20.49 | 3.86 | 29.60 |
| extra over for | | | | | | |
| offset, 75mm | nr | 0.30 | 5.25 | 38.72 | 6.60 | 50.57 |
| offset, 100mm | nr | 0.30 | 5.25 | 38.72 | 6.60 | 50.57 |
| offset, 150mm | nr | 0.30 | 5.25 | 38.72 | 6.60 | 50.57 |
| offset, 225mm | nr | 0.30 | 5.25 | 45.86 | 7.67 | 58.78 |
| offset, 300mm | nr | 0.30 | 5.25 | 46.85 | 7.82 | 59.92 |
| bend, 92.5 degrees | nr | 0.30 | 5.25 | 28.79 | 5.11 | 39.15 |
| branch | nr | 0.30 | 5.25 | 32.92 | 5.73 | 43.90 |
| shoe | nr | 0.30 | 5.25 | 13.19 | 2.77 | 21.21 |
| standard hopper | nr | 0.30 | 5.25 | 27.73 | 4.95 | 37.93 |
| flatback hopper | nr | 0.30 | 5.25 | 18.59 | 3.58 | 27.42 |
| 102mm diameter pipe | m | 0.32 | 5.60 | 28.62 | 5.13 | 39.35 |
| extra over for | | | | | | |
| offset, 75mm | nr | 0.32 | 5.60 | 47.07 | 7.90 | 60.57 |
| offset, 100mm | nr | 0.32 | 5.60 | 47.07 | 7.90 | 60.57 |
| offset, 150mm | nr | 0.32 | 5.60 | 47.07 | 7.90 | 60.57 |
| offset, 225mm | nr | 0.32 | 5.60 | 54.93 | 9.08 | 69.61 |
| offset, 300mm | nr | 0.32 | 5.60 | 54.93 | 9.08 | 69.61 |
| bend, 92.5 degrees | nr | 0.32 | 5.60 | 41.28 | 7.03 | 53.91 |
| branch | nr | 0.32 | 5.60 | 44.27 | 7.48 | 57.35 |
| shoe | nr | 0.32 | 5.60 | 17.41 | 3.45 | 26.46 |
| standard hopper | nr | 0.32 | 5.60 | 37.26 | 6.43 | 49.29 |
| ornamental hopper | nr | 0.32 | 5.60 | 130.20 | 20.37 | 156.17 |

| | Unit | Labour hours | Labour cost £ | Materials £ | O & P £ | Total £ |
|---|---|---|---|---|---|---|
| Cast iron rainwater pipe, straight, ears cast on, spigot and socket dry joints plugged to brickwork | | | | | | |
| | | | | | | |
| 65mm diameter pipe | m | 0.25 | 4.38 | 21.20 | 3.84 | 29.41 |
| extra over for | | | | | | |
| offset, 75mm | nr | 0.25 | 4.38 | 16.57 | 3.14 | 24.09 |
| offset, 115mm | nr | 0.25 | 4.38 | 16.57 | 3.14 | 24.09 |
| offset, 150mm | nr | 0.25 | 4.38 | 16.57 | 3.14 | 24.09 |
| offset, 225mm | nr | 0.25 | 4.38 | 19.30 | 3.55 | 27.23 |
| offset, 305mm | nr | 0.25 | 4.38 | 22.60 | 4.05 | 31.02 |
| bend | nr | 0.25 | 4.38 | 10.82 | 2.28 | 17.47 |
| branch | nr | 0.25 | 4.38 | 21.28 | 3.85 | 29.50 |
| shoe | nr | 0.25 | 4.38 | 15.34 | 2.96 | 22.67 |
| eared shoe | nr | 0.25 | 4.38 | 17.68 | 3.31 | 25.36 |
| flat hopper | nr | 0.25 | 4.38 | 13.82 | 2.73 | 20.92 |
| rectangular hopper | nr | 0.25 | 4.38 | 61.66 | 9.91 | 75.94 |
| | | | | | | |
| 75mm diameter pipe | m | 0.30 | 5.25 | 22.25 | 4.13 | 31.63 |
| extra over for | | | | | | |
| offset, 75mm | nr | 0.30 | 5.25 | 16.57 | 3.27 | 25.09 |
| offset, 115mm | nr | 0.30 | 5.25 | 16.57 | 3.27 | 25.09 |
| offset, 150mm | nr | 0.30 | 5.25 | 16.57 | 3.27 | 25.09 |
| offset, 225mm | nr | 0.30 | 5.25 | 19.30 | 3.68 | 28.23 |
| offset, 305mm | nr | 0.30 | 5.25 | 23.71 | 4.34 | 33.30 |
| bend | nr | 0.30 | 5.25 | 13.15 | 2.76 | 21.16 |
| branch | nr | 0.30 | 5.25 | 23.02 | 4.24 | 32.51 |
| shoe | nr | 0.30 | 5.25 | 15.34 | 3.09 | 23.68 |
| eared shoe | nr | 0.30 | 5.25 | 17.68 | 3.44 | 26.37 |
| flat hopper | nr | 0.30 | 5.25 | 15.69 | 3.14 | 24.08 |
| rectangular hopper | nr | 0.30 | 5.25 | 61.56 | 10.02 | 76.83 |
| | | | | | | |
| 100mm diameter pipe | m | 0.35 | 6.13 | 29.07 | 5.28 | 40.47 |
| extra over for | | | | | | |
| offset, 75mm | nr | 0.35 | 6.13 | 31.28 | 5.61 | 43.02 |
| offset, 115mm | nr | 0.35 | 6.13 | 31.88 | 5.70 | 43.71 |
| offset, 150mm | nr | 0.35 | 6.13 | 31.28 | 5.61 | 43.02 |
| offset, 225mm | nr | 0.35 | 6.13 | 37.87 | 6.60 | 50.59 |

| | Unit | Labour hours | Labour cost £ | Materials £ | O & P £ | Total £ |
|---|---|---|---|---|---|---|
| offset, 305mm | nr | 0.35 | 6.13 | 61.56 | 10.15 | 77.84 |
| bend | nr | 0.35 | 6.13 | 18.57 | 3.70 | 28.40 |
| branch | nr | 0.35 | 6.13 | 27.35 | 5.02 | 38.50 |
| shoe | nr | 0.35 | 6.13 | 20.68 | 4.02 | 30.83 |
| eared shoe | nr | 0.35 | 6.13 | 23.50 | 4.44 | 34.07 |
| flat hopper | nr | 0.35 | 6.13 | 34.78 | 6.14 | 47.04 |
| rectangular hopper | nr | 0.35 | 6.13 | 61.56 | 10.15 | 77.84 |

PVC-U half round rainwater
gutter, fixed to timber with
support brackets at 1m
maximum centres

| | Unit | Labour hours | Labour cost £ | Materials £ | O & P £ | Total £ |
|---|---|---|---|---|---|---|
| 76mm elliptical gutter | m | 0.20 | 3.50 | 6.58 | 1.51 | 11.59 |
| extra over for | | | | | | |
| running outlet | nr | 0.20 | 3.50 | 6.00 | 1.43 | 10.93 |
| angle | nr | 0.20 | 3.50 | 6.54 | 1.51 | 11.55 |
| stop end outlet | nr | 0.10 | 1.75 | 5.26 | 1.05 | 8.06 |
| stop end | nr | 0.10 | 1.75 | 3.22 | 0.75 | 5.72 |
| 112mm half-round gutter | m | 0.26 | 4.55 | 4.86 | 1.41 | 10.82 |
| extra over for | | | | | | |
| running outlet | nr | 0.26 | 4.55 | 4.23 | 1.32 | 10.10 |
| angle | nr | 0.26 | 4.55 | 4.82 | 1.41 | 10.78 |
| stop end outlet | nr | 0.13 | 2.28 | 4.23 | 0.98 | 7.48 |
| stop end | nr | 0.13 | 2.28 | 2.17 | 0.67 | 5.11 |
| 150mm half-round gutter | m | 0.30 | 5.25 | 7.17 | 1.86 | 14.28 |
| extra over for | | | | | | |
| running outlet | nr | 0.30 | 5.25 | 9.76 | 2.25 | 17.26 |
| angle | nr | 0.30 | 5.25 | 4.81 | 1.51 | 11.57 |
| stop end outlet | nr | 0.15 | 2.63 | 6.14 | 1.31 | 10.08 |
| stop end | nr | 0.15 | 2.63 | 1.97 | 0.69 | 5.28 |

| | Unit | Labour hours | Labour cost £ | Materials £ | O & P £ | Total £ |
|---|---|---|---|---|---|---|
| **Aluminium half round rainwater gutter with mastic joints, fixed to timber with support brackets at 1m maximum centres** | | | | | | |
| 100mm gutter | m | 0.30 | 5.25 | 13.33 | 2.79 | 21.37 |
| extra over for | | | | | | |
| running outlet | nr | 0.30 | 5.25 | 11.89 | 2.57 | 19.71 |
| angle | nr | 0.30 | 5.25 | 6.99 | 1.84 | 14.08 |
| stop end outlet | nr | 0.15 | 2.63 | 6.83 | 1.42 | 10.87 |
| stop end | nr | 0.15 | 2.63 | 2.63 | 0.79 | 6.04 |
| 125mm gutter | m | 0.32 | 5.60 | 16.80 | 3.36 | 25.76 |
| extra over for | | | | | | |
| running outlet | nr | 0.32 | 5.60 | 11.37 | 2.55 | 19.52 |
| angle | nr | 0.32 | 5.60 | 9.53 | 2.27 | 17.40 |
| stop end outlet | nr | 0.16 | 2.80 | 7.86 | 1.60 | 12.26 |
| stop end | nr | 0.16 | 2.80 | 3.42 | 0.93 | 7.15 |
| **Aluminium ogee rainwater gutter with mastic joints, fixed to timber with support brackets at 1m maximum centres** | | | | | | |
| 120 × 75mm gutter | m | 0.30 | 5.25 | 17.22 | 3.37 | 25.84 |
| extra over for | | | | | | |
| running outlet | nr | 0.30 | 5.25 | 21.70 | 4.04 | 30.99 |
| angle | nr | 0.30 | 5.25 | 18.17 | 3.51 | 26.93 |
| stop end outlet | nr | 0.15 | 2.63 | 30.32 | 4.94 | 37.89 |
| stop end | nr | 0.15 | 2.63 | 4.23 | 1.03 | 7.88 |
| 155 × 100mm gutter | m | 0.32 | 5.60 | 32.49 | 5.71 | 43.80 |
| extra over for | | | | | | |
| running outlet | nr | 0.32 | 5.60 | 28.69 | 5.14 | 39.43 |
| angle | nr | 0.32 | 5.60 | 28.05 | 5.05 | 38.70 |
| stop end outlet | nr | 0.16 | 2.80 | 23.38 | 3.93 | 30.11 |
| stop end | nr | 0.16 | 2.80 | 17.07 | 2.98 | 22.85 |

|  | Unit | Labour hours | Labour cost £ | Materials £ | O & P £ | Total £ |
|---|---|---|---|---|---|---|
| Cast iron half-round rainwater gutter with mastic joints, primed, fixed to timber with support brackets at 1m maximum centres | | | | | | |
| 100mm gutter | m | 0.36 | 6.30 | 11.77 | 2.71 | 20.78 |
| extra over for | | | | | | |
| running outlet | nr | 0.36 | 6.30 | 7.56 | 2.08 | 15.94 |
| angle | nr | 0.36 | 6.30 | 7.76 | 2.11 | 16.17 |
| stop end outlet | nr | 0.18 | 3.15 | 8.87 | 1.80 | 13.82 |
| stop end | nr | 0.18 | 3.15 | 2.61 | 0.86 | 6.62 |
| 115mm gutter | m | 0.38 | 6.65 | 12.15 | 2.82 | 21.62 |
| extra over for | | | | | | |
| running outlet | nr | 0.38 | 6.65 | 8.24 | 2.23 | 17.12 |
| angle | nr | 0.38 | 6.65 | 7.99 | 2.20 | 16.84 |
| stop end outlet | nr | 0.19 | 3.33 | 9.94 | 1.99 | 15.25 |
| stop end | nr | 0.19 | 3.33 | 3.38 | 1.01 | 7.71 |
| 125mm gutter | m | 0.40 | 7.00 | 14.05 | 3.16 | 24.21 |
| extra over for | | | | | | |
| running outlet | nr | 0.40 | 7.00 | 7.84 | 2.23 | 17.07 |
| angle | nr | 0.40 | 7.00 | 9.42 | 2.46 | 18.88 |
| stop end outlet | nr | 0.20 | 3.50 | 11.34 | 2.23 | 17.07 |
| stop end | nr | 0.20 | 3.50 | 3.38 | 1.03 | 7.91 |
| 150mm gutter | m | 0.44 | 7.70 | 23.47 | 4.68 | 35.85 |
| extra over for | | | | | | |
| running outlet | nr | 0.44 | 7.70 | 14.71 | 3.36 | 25.77 |
| angle | nr | 0.44 | 7.70 | 17.21 | 3.74 | 28.65 |
| stop end outlet | nr | 0.22 | 3.85 | 18.66 | 3.38 | 25.89 |
| stop end | nr | 0.22 | 3.85 | 4.68 | 1.28 | 9.81 |

| | Unit | Labour hours | Labour cost £ | Materials £ | O & P £ | Total £ |
|---|---|---|---|---|---|---|
| **Roof outlets** | | | | | | |
| | | | | | | |
| Cast iron circular roof outlet with flat grate, diameter | | | | | | |
| 50mm | nr | 0.50 | 8.75 | 81.01 | 13.46 | 103.22 |
| 75mm | nr | 0.60 | 10.50 | 88.35 | 14.83 | 113.68 |
| 100mm | nr | 0.70 | 12.25 | 103.44 | 17.35 | 133.04 |
| | | | | | | |
| Aluminium circular roof outlet with domed grate, diameter | | | | | | |
| 50mm | nr | 0.50 | 8.75 | 74.70 | 12.52 | 95.97 |
| 75mm | nr | 0.60 | 10.50 | 85.28 | 14.37 | 110.15 |
| 100mm | nr | 0.70 | 12.25 | 110.08 | 18.35 | 140.68 |
| 150mm | nr | 0.80 | 14.00 | 123.20 | 20.58 | 157.78 |
| | | | | | | |
| Plastic wire balloon guard for pipes and outlets, diameter | | | | | | |
| 50mm | nr | 0.05 | 0.88 | 2.14 | 0.45 | 3.47 |
| 63mm | nr | 0.05 | 0.88 | 2.27 | 0.47 | 3.62 |
| 75mm | nr | 0.05 | 0.88 | 2.34 | 0.48 | 3.70 |
| 100mm | nr | 0.05 | 0.88 | 2.36 | 0.49 | 3.72 |
| | | | | | | |
| Galvanised wire balloon guard for pipes and outlets, diameter | | | | | | |
| 50mm | nr | 0.05 | 0.88 | 2.79 | 0.55 | 4.21 |
| 63mm | nr | 0.05 | 0.88 | 1.85 | 0.41 | 3.13 |
| 75mm | nr | 0.05 | 0.88 | 1.89 | 0.41 | 3.18 |
| 100mm | nr | 0.05 | 0.88 | 2.36 | 0.49 | 3.72 |
| | | | | | | |
| Copper wire balloon guard for pipes and outlets, diameter | | | | | | |
| 50mm | nr | 0.05 | 0.88 | 2.13 | 0.45 | 3.46 |
| 63mm | nr | 0.05 | 0.88 | 2.17 | 0.46 | 3.50 |
| 75mm | nr | 0.05 | 0.88 | 2.33 | 0.48 | 3.69 |
| 100mm | nr | 0.05 | 0.88 | 3.27 | 0.62 | 4.77 |

| | Unit | Labour hours | Labour cost £ | Materials £ | O & P £ | Total £ |
|---|---|---|---|---|---|---|

**Drainage above ground level**

**Waste systems**

MuPVC waste system, solvent welded joints, clips at 500mm maximum centres, plugged to brickwork

| | Unit | Labour hours | Labour cost £ | Materials £ | O & P £ | Total £ |
|---|---|---|---|---|---|---|
| 32mm diameter pipe | m | 0.25 | 4.38 | 3.29 | 1.15 | 8.81 |
| extra over for | | | | | | |
| bend, 45° | nr | 0.24 | 4.20 | 2.00 | 0.93 | 7.13 |
| bend, 87.5° | nr | 0.24 | 4.20 | 2.00 | 0.93 | 7.13 |
| long tail bend, 90° | nr | 0.24 | 4.20 | 2.00 | 0.93 | 7.13 |
| tee | nr | 0.24 | 4.20 | 3.42 | 1.14 | 8.76 |
| bottle trap | nr | 0.24 | 4.20 | 4.65 | 1.33 | 10.18 |
| tubular P trap | nr | 0.24 | 4.20 | 4.57 | 1.32 | 10.09 |
| tubular S trap | nr | 0.24 | 4.20 | 5.48 | 1.45 | 11.13 |
| 40mm diameter pipe | m | 0.28 | 4.90 | 3.54 | 1.27 | 9.71 |
| extra over for | | | | | | |
| bend, 45° | nr | 0.26 | 4.55 | 2.29 | 1.03 | 7.87 |
| bend, 87.5° | nr | 0.26 | 4.55 | 2.29 | 1.03 | 7.87 |
| long tail bend, 90° | nr | 0.26 | 4.55 | 2.29 | 1.03 | 7.87 |
| tee | nr | 0.26 | 4.55 | 4.12 | 1.30 | 9.97 |
| cross tee | nr | 0.26 | 4.55 | 11.03 | 2.34 | 17.92 |
| bottle trap | nr | 0.26 | 4.55 | 5.28 | 1.47 | 11.30 |
| tubular P trap | nr | 0.26 | 4.55 | 5.48 | 1.50 | 11.53 |
| tubular S trap | nr | 0.26 | 4.55 | 6.52 | 1.66 | 12.73 |
| running P trap | nr | 0.26 | 4.55 | 7.84 | 1.86 | 14.25 |
| 50mm diameter pipe | m | 0.34 | 5.95 | 4.98 | 1.64 | 12.57 |
| extra over for | | | | | | |
| bend, 45° | nr | 0.32 | 5.60 | 3.21 | 1.32 | 10.13 |
| bend, 87.5° | nr | 0.32 | 5.60 | 3.21 | 1.32 | 10.13 |
| long tail bend, 90° | nr | 0.32 | 5.60 | 3.21 | 1.32 | 10.13 |
| tee | nr | 0.32 | 5.60 | 6.91 | 1.88 | 14.39 |
| cross tee | nr | 0.32 | 5.60 | 12.79 | 2.76 | 21.15 |

| | Unit | Labour hours | Labour cost £ | Materials £ | O & P £ | Total £ |
|---|---|---|---|---|---|---|
| **MuPVC waste system, push-fit joints, clips at 500mm maximum centres, plugged to brickwork** | | | | | | |
| 32mm diameter pipe | m | 0.20 | 3.50 | 1.18 | 0.70 | 5.38 |
| extra over for | | | | | | |
| bend, 45° | nr | 0.18 | 3.15 | 1.00 | 0.62 | 4.77 |
| knuckle bend, 90° | nr | 0.18 | 3.15 | 0.98 | 0.62 | 4.75 |
| spigot bend, 90° | nr | 0.18 | 3.15 | 1.23 | 0.66 | 5.04 |
| tee | nr | 0.18 | 3.15 | 1.56 | 0.71 | 5.42 |
| P trap, 38mm | nr | 0.18 | 3.15 | 4.57 | 1.16 | 8.88 |
| bottle trap, 38mm | nr | 0.18 | 3.15 | 4.65 | 1.17 | 8.97 |
| anti-syphon bottle trap, 76mm | nr | 0.18 | 3.15 | 7.43 | 1.59 | 12.17 |
| connection to back inlet gulley, caulking bush | nr | 0.16 | 2.80 | 2.29 | 0.76 | 5.85 |
| 40mm diameter pipe | m | 0.24 | 4.20 | 2.00 | 0.93 | 7.13 |
| extra over for | | | | | | |
| bend, 45° | nr | 0.20 | 3.50 | 1.06 | 0.68 | 5.24 |
| knuckle bend, 90° | nr | 0.20 | 3.50 | 1.00 | 0.68 | 5.18 |
| spigot bend, 90° | nr | 0.20 | 3.50 | 1.29 | 0.72 | 5.51 |
| tee | nr | 0.20 | 3.50 | 1.61 | 0.77 | 5.88 |
| P trap, 38mm | nr | 0.20 | 3.50 | 5.48 | 1.35 | 10.33 |
| bottle trap, 38mm | nr | 0.20 | 3.50 | 5.28 | 1.32 | 10.10 |
| anti-syphon bottle trap, 76mm | nr | 0.20 | 3.50 | 8.60 | 1.82 | 13.92 |
| connection to back inlet gulley, caulking bush | nr | 0.18 | 3.15 | 3.12 | 0.94 | 7.21 |
| **PVC-U soil system, solvent-welded joints, holderbats at 1250mm maximum centres, plugged to brickwork** | | | | | | |
| 82mm diameter pipe | m | 0.34 | 5.95 | 13.49 | 2.92 | 22.36 |
| extra over for | | | | | | |
| bend, 92.5° | nr | 0.30 | 5.25 | 13.60 | 2.83 | 21.68 |
| bend, 135° | nr | 0.30 | 5.25 | 13.60 | 2.83 | 21.68 |

| | Unit | Labour hours | Labour cost £ | Materials £ | O & P £ | Total £ |
|---|---|---|---|---|---|---|
| bend, spigot/spigot | nr | 0.30 | 5.25 | 13.60 | 2.83 | 21.68 |
| branch, single, 92.5° | nr | 0.38 | 6.65 | 20.00 | 4.00 | 30.65 |
| branch, single, 135° | nr | 0.38 | 6.65 | 20.00 | 4.00 | 30.65 |
| access pipe, single branch | nr | 0.38 | 6.65 | 20.00 | 4.00 | 30.65 |
| access cap | nr | 0.22 | 3.85 | 12.27 | 2.42 | 18.54 |
| vent cowl | nr | 0.22 | 3.85 | 3.68 | 1.13 | 8.66 |
| 110mm diameter pipe | m | 0.38 | 6.65 | 11.53 | 2.73 | 20.91 |
| extra over for | | | | | | |
| bend, 92.5° | nr | 0.34 | 5.95 | 16.15 | 3.32 | 25.42 |
| bend, 135° | nr | 0.34 | 5.95 | 16.15 | 3.32 | 25.42 |
| bend, variable | nr | 0.34 | 5.95 | 16.15 | 3.32 | 25.42 |
| bend, spigot/spigot | nr | 0.34 | 5.95 | 14.82 | 3.12 | 23.89 |
| branch, single, 92.5° | nr | 0.40 | 7.00 | 21.04 | 4.21 | 32.25 |
| branch, single, 135° | nr | 0.40 | 7.00 | 21.04 | 4.21 | 32.25 |
| branch, single, 92.5° spigot outlet | nr | 0.40 | 7.00 | 21.04 | 4.21 | 32.25 |
| access pipe connector | nr | 0.34 | 5.95 | 21.58 | 4.13 | 31.66 |
| access pipe, single branch | nr | 0.34 | 5.95 | 21.04 | 4.05 | 31.04 |
| access cap | nr | 0.38 | 6.65 | 12.96 | 2.94 | 22.55 |
| WC manifold connector | nr | 0.38 | 6.65 | 9.88 | 2.48 | 19.01 |
| vent cowl | nr | 0.22 | 3.85 | 3.88 | 1.16 | 8.89 |
| 150mm diameter pipe | m | 0.42 | 7.35 | 31.32 | 5.80 | 44.47 |
| extra over for | | | | | | |
| bend, 92.5° | nr | 0.38 | 6.65 | 43.88 | 7.58 | 58.11 |
| bend, 135° | nr | 0.38 | 6.65 | 43.88 | 7.58 | 58.11 |
| bend, spigot/spigot | nr | 0.38 | 6.65 | 43.88 | 7.58 | 58.11 |
| bend, spigot/socket | nr | 0.38 | 6.65 | 43.88 | 7.58 | 58.11 |
| branch, single, 92.5° | nr | 0.42 | 7.35 | 72.27 | 11.94 | 91.56 |
| branch, single, 135° | nr | 0.42 | 7.35 | 72.27 | 11.94 | 91.56 |
| access cap | nr | 0.40 | 7.00 | 35.55 | 6.38 | 48.93 |
| vent cowl | nr | 0.24 | 4.20 | 10.94 | 2.27 | 17.41 |

| | Unit | Labour hours | Labour cost £ | Materials £ | O & P £ | Total £ |
|---|---|---|---|---|---|---|

**Soil pipes**

Cast iron soil system with
flexible joints, pipe brackets at
2m maximum centres, plugged
to brickwork

| | Unit | Labour hours | Labour cost £ | Materials £ | O & P £ | Total £ |
|---|---|---|---|---|---|---|
| 50mm diameter pipe | m | 0.55 | 9.63 | 21.66 | 4.69 | 35.98 |
| extra over for | | | | | | |
| bend, short radius | nr | 0.45 | 7.88 | 10.72 | 2.79 | 21.38 |
| bend, short radius with | | | | | | |
| door | nr | 0.45 | 7.88 | 26.44 | 5.15 | 39.46 |
| branch, single, plain | nr | 0.50 | 8.75 | 16.14 | 3.73 | 28.62 |
| branch, single with door | nr | 0.50 | 8.75 | 31.86 | 6.09 | 46.70 |
| branch, double, plain | nr | 0.50 | 8.75 | 28.39 | 5.57 | 42.71 |
| P trap with door | nr | 0.45 | 7.88 | 34.67 | 6.38 | 48.93 |
| 75mm diameter pipe | m | 0.60 | 10.50 | 21.44 | 4.79 | 36.73 |
| extra over for | | | | | | |
| bend, short radius | nr | 0.50 | 8.75 | 10.72 | 2.92 | 22.39 |
| bend, short radius with | | | | | | |
| door | nr | 0.50 | 8.75 | 26.44 | 5.28 | 40.47 |
| bend, long radius | nr | 0.50 | 8.75 | 21.26 | 4.50 | 34.51 |
| bend, long radius, | | | | | | |
| access | nr | 0.50 | 8.75 | 26.44 | 5.28 | 40.47 |
| branch, single, plain | nr | 0.55 | 9.63 | 16.14 | 3.86 | 29.63 |
| branch, single with door | nr | 0.55 | 9.63 | 31.86 | 6.22 | 47.71 |
| branch, double, plain | nr | 0.55 | 9.63 | 27.15 | 5.52 | 42.29 |
| offset, 75mm projection | nr | 0.50 | 8.75 | 10.59 | 2.90 | 22.24 |
| offset, 115mm | | | | | | |
| projection | nr | 0.50 | 8.75 | 13.19 | 3.29 | 25.23 |
| offset, 150mm | | | | | | |
| projection | nr | 0.50 | 8.75 | 13.19 | 3.29 | 25.23 |
| offset, 225mm | | | | | | |
| projection | nr | 0.50 | 8.75 | 16.53 | 3.79 | 29.07 |
| offset, 300mm | nr | 0.50 | 8.75 | 19.50 | 4.24 | 32.49 |
| projection | | | | | | |
| blank end, plain | nr | 0.50 | 8.75 | 3.85 | 1.89 | 14.49 |
| blank end, drilled and | | | | | | |
| tapped | nr | 0.50 | 8.75 | 9.15 | 2.69 | 20.59 |
| P trap, plain | nr | 0.50 | 8.75 | 20.68 | 4.41 | 33.84 |
| P trap with door | nr | 0.50 | 8.75 | 36.40 | 6.77 | 51.92 |

| | Unit | Labour hours | Labour cost £ | Materials £ | O & P £ | Total £ |
|---|---|---|---|---|---|---|
| WC connector, 305mm effective length | nr | 0.50 | 8.75 | 19.80 | 4.28 | 32.83 |
| 100mm diameter pipe | m | 0.65 | 11.38 | 32.12 | 6.52 | 50.02 |
| extra over for | | | | | | |
| bend, short radius | nr | 0.55 | 9.63 | 14.84 | 3.67 | 28.13 |
| bend, short radius with door | nr | 0.55 | 9.63 | 31.39 | 6.15 | 47.17 |
| bend, long radius | nr | 0.55 | 9.63 | 24.04 | 5.05 | 38.71 |
| bend, long radius, access | nr | 0.55 | 9.63 | 40.61 | 7.54 | 57.77 |
| bend, long radius, heel rest | nr | 0.55 | 9.63 | 29.11 | 5.81 | 44.55 |
| bend, long tail, 87.5° | nr | 0.55 | 9.63 | 19.18 | 4.32 | 33.13 |
| branch, single, plain | nr | 0.60 | 10.50 | 22.94 | 5.02 | 38.46 |
| branch, single with door | nr | 0.60 | 10.50 | 39.53 | 7.50 | 57.53 |
| branch, double, plain | nr | 0.60 | 10.50 | 28.39 | 5.83 | 44.72 |
| branch, double with door | nr | 0.60 | 10.50 | 44.95 | 8.32 | 63.77 |
| branch, corner | nr | 0.60 | 10.50 | 37.35 | 7.18 | 55.03 |
| offset, 75mm projection | nr | 0.55 | 9.63 | 15.61 | 3.79 | 29.02 |
| offset, 115mm projection | nr | 0.55 | 9.63 | 18.61 | 4.24 | 32.47 |
| offset, 150mm projection | nr | 0.55 | 9.63 | 18.61 | 4.24 | 32.47 |
| offset, 225mm projection | nr | 0.55 | 9.63 | 21.34 | 4.64 | 35.61 |
| offset, 300mm projection | nr | 0.55 | 9.63 | 24.04 | 5.05 | 38.71 |
| blank end, plain | nr | 0.55 | 9.63 | 4.52 | 2.12 | 16.27 |
| blank end, drilled and tapped | nr | 0.55 | 9.63 | 9.83 | 2.92 | 22.37 |
| P trap, plain | nr | 0.55 | 9.63 | 23.79 | 5.01 | 38.43 |
| P trap with door | nr | 0.55 | 9.63 | 40.36 | 7.50 | 57.48 |
| WC connector, 305mm effective length | nr | 0.55 | 9.63 | 13.50 | 3.47 | 26.59 |

| | Unit | Labour hours | Labour cost £ | Materials £ | O & P £ | Total £ |
|---|---|---|---|---|---|---|
| 150mm diameter pipe | m | 0.70 | 12.25 | 62.38 | 11.19 | 85.82 |
| extra over for | | | | | | |
| bend, short radius | nr | 0.60 | 10.50 | 26.51 | 5.55 | 42.56 |
| bend, short radius with door | nr | 0.60 | 10.50 | 44.60 | 8.27 | 63.37 |
| bend, long radius | nr | 0.60 | 10.50 | 43.53 | 8.10 | 62.13 |
| bend, long radius, access | nr | 0.60 | 10.50 | 6.15 | 2.50 | 19.15 |
| branch, single, plain | nr | 0.65 | 11.38 | 47.20 | 8.79 | 67.36 |
| branch, single with door | nr | 0.65 | 11.38 | 65.34 | 11.51 | 88.22 |
| branch, double, plain | nr | 0.65 | 11.38 | 70.13 | 12.23 | 93.73 |
| blank end, plain | nr | 0.60 | 10.50 | 6.52 | 2.55 | 19.57 |
| blank end, drilled and tapped | nr | 0.60 | 10.50 | 118.81 | 19.40 | 148.71 |
| P trap with door | nr | 0.60 | 10.50 | 70.28 | 12.12 | 92.90 |

**Overflow systems**

MuPVC overflow system, solvent-welded joints, clips at 500mm maximum centres, plugged to brickwork

| | Unit | Labour hours | Labour cost £ | Materials £ | O & P £ | Total £ |
|---|---|---|---|---|---|---|
| 19mm diameter pipe | m | 0.20 | 3.50 | 1.87 | 0.81 | 6.18 |
| extra over for | | | | | | |
| bend | nr | 0.18 | 3.15 | 1.37 | 0.68 | 5.20 |
| tee | nr | 0.18 | 3.15 | 1.42 | 0.69 | 5.26 |
| connector, bent tank | nr | 0.20 | 3.50 | 2.03 | 0.83 | 6.36 |

**Traps**

Polypropylene traps, screwed joints to outlet and pipe

| | Unit | Labour hours | Labour cost £ | Materials £ | O & P £ | Total £ |
|---|---|---|---|---|---|---|
| Bottle trap, 38mm seal | | | | | | |
| 32mm | nr | 0.30 | 5.25 | 4.65 | 1.49 | 11.39 |
| 40mm | nr | 0.35 | 6.13 | 4.74 | 1.63 | 12.49 |
| Bottle trap, 76mm seal | | | | | | |
| 32mm | nr | 0.30 | 5.25 | 4.65 | 1.49 | 11.39 |
| 40mm | nr | 0.35 | 6.13 | 4.74 | 1.63 | 12.49 |
| Anti-syphon bottle trap | | | | | | |
| 32mm | nr | 0.30 | 5.25 | 7.43 | 1.90 | 14.58 |
| 40mm | nr | 0.35 | 6.13 | 7.72 | 2.08 | 15.92 |

|  | Unit | Labour hours | Labour cost £ | Materials £ | O & P £ | Total £ |
|---|---|---|---|---|---|---|
| Tubular S trap |  |  |  |  |  |  |
| 32mm | nr | 0.30 | 5.25 | 5.48 | 1.61 | 12.34 |
| 40mm | nr | 0.35 | 6.13 | 6.52 | 1.90 | 14.54 |
| Tubular P trap |  |  |  |  |  |  |
| 32mm | nr | 0.30 | 5.25 | 4.57 | 1.47 | 11.29 |
| 40mm | nr | 0.35 | 6.13 | 5.48 | 1.74 | 13.35 |
| Running P trap |  |  |  |  |  |  |
| 40mm | nr | 0.35 | 6.13 | 7.84 | 2.09 | 16.06 |
| Bath trap |  |  |  |  |  |  |
| 40mm | nr | 0.35 | 6.13 | 6.88 | 1.95 | 14.96 |
| Bath trap with overflow |  |  |  |  |  |  |
| 40mm | nr | 0.35 | 6.13 | 6.43 | 1.88 | 14.44 |
| Washing machine, half trap |  |  |  |  |  |  |
| 40mm | nr | 0.35 | 6.13 | 4.57 | 1.60 | 12.30 |
| **Copper pipework, capillary joints** |  |  |  |  |  |  |
| 15mm diameter, to timber | m | 0.20 | 3.50 | 1.85 | 0.80 | 6.15 |
| 15mm diameter, plugged and screwed | m | 0.22 | 3.85 | 2.05 | 0.89 | 6.79 |
| extra over for |  |  |  |  |  |  |
| reduced coupling, 15 × 8mm | nr | 0.18 | 3.15 | 3.77 | 1.04 | 7.96 |
| reduced coupling, 15 × 10mm | nr | 0.18 | 3.15 | 1.50 | 0.70 | 5.35 |
| reduced coupling, 15 × 12mm | nr | 0.18 | 3.15 | 2.93 | 0.91 | 6.99 |

|  | Unit | Labour hours | Labour cost £ | Materials £ | O & P £ | Total £ |
|---|---|---|---|---|---|---|
| **Extra over 15mm diameter pipework (cont'd)** | | | | | | |
| adaptor coupling, 15mm × 1/2in | nr | 0.18 | 3.15 | 9.58 | 1.91 | 14.64 |
| straight female coupling, 15mm × 1/2in | nr | 0.18 | 3.15 | 5.57 | 1.31 | 10.03 |
| female reducing connector, 15mm × 3/8in | nr | 0.18 | 3.15 | 11.07 | 2.13 | 16.35 |
| female reducing connector, 15mm × 1/4in | nr | 0.18 | 3.15 | 9.87 | 1.95 | 14.97 |
| straight male coupling, 15mm × 1/2in | nr | 0.18 | 3.15 | 4.74 | 1.18 | 9.07 |
| male reducing connector, 15mm × 3/4in | nr | 0.18 | 3.15 | 12.33 | 2.32 | 17.80 |
| tank connector, 15mm × 1/2in | nr | 0.18 | 3.15 | 13.03 | 2.43 | 18.61 |
| reducer, 15 × 8mm | nr | 0.18 | 3.15 | 3.77 | 1.04 | 7.96 |
| reducer, 15 × 10mm | nr | 0.18 | 3.15 | 1.50 | 0.70 | 5.35 |
| reducer, 15 × 12mm | nr | 0.18 | 3.15 | 2.93 | 0.91 | 6.99 |
| female adaptor, 15 × 1/2in | nr | 0.18 | 3.15 | 9.58 | 1.91 | 14.64 |
| male adaptor, 15 × 1/2in | nr | 0.18 | 3.15 | 9.78 | 1.94 | 14.87 |
| adaptor, 15 × 1/2in | nr | 0.18 | 3.15 | 4.65 | 1.17 | 8.97 |
| male elbow, 15mm × 1/2in | nr | 0.18 | 3.15 | 9.48 | 1.89 | 14.52 |
| reducing male elbow, 15mm × 3/8in | nr | 0.18 | 3.15 | 13.36 | 2.48 | 18.99 |
| female elbow, 15mm × 1/2in | nr | 0.18 | 3.15 | 8.18 | 1.70 | 13.03 |
| reducing female elbow, 15mm × 1/4in | nr | 0.18 | 3.15 | 13.04 | 2.43 | 18.62 |
| backplate elbow, 15mm × 1/2in | nr | 0.18 | 3.15 | 21.04 | 3.63 | 27.82 |
| flanged bend, 15mm × 1/2in | nr | 0.18 | 3.15 | 21.58 | 3.71 | 28.44 |
| flanged bend, 15mm × 3/4in | nr | 0.18 | 3.15 | 22.27 | 3.81 | 29.23 |
| slow bend | nr | 0.18 | 3.15 | 1.61 | 0.71 | 5.47 |
| return bend | nr | 0.18 | 3.15 | 14.88 | 2.70 | 20.73 |
| obtuse elbow | nr | 0.18 | 3.15 | 1.89 | 0.76 | 5.80 |
| tee, reduced branch largest end 15mm | nr | 0.22 | 3.85 | 16.74 | 3.09 | 23.68 |
| tee, both ends reduced, largest end 15mm | nr | 0.22 | 3.85 | 16.74 | 3.09 | 23.68 |

| | Unit | Labour hours | Labour cost £ | Materials £ | O & P £ | Total £ |
|---|---|---|---|---|---|---|
| sweep tee, 90 degrees, 15mm | nr | 0.22 | 3.85 | 16.78 | 3.09 | 23.72 |
| offset tee, 15mm | nr | 0.22 | 3.85 | 22.21 | 3.91 | 29.97 |
| double sweep tee, 15mm | nr | 0.22 | 3.85 | 18.98 | 3.42 | 26.25 |
| cross equal tee, 15mm | nr | 0.22 | 3.85 | 1.33 | 0.78 | 5.96 |
| stop end, 15mm | nr | 0.18 | 3.15 | 2.44 | 0.84 | 6.43 |
| tap connector, 15mm | nr | 0.18 | 3.15 | 3.43 | 0.99 | 7.57 |
| bent union adaptor, 15mm × 3/4in | nr | 0.18 | 3.15 | 4.65 | 1.17 | 8.97 |
| bent male union connector, 15mm × 1/2in | nr | 0.18 | 3.15 | 4.74 | 1.18 | 9.07 |
| bent female union connector, 15mm × 1/2in | nr | 0.18 | 3.15 | 5.57 | 1.31 | 10.03 |
| made bend | nr | 0.18 | 3.15 | 0.00 | 0.47 | 3.62 |
| 22mm diameter, to timber | m | 0.22 | 3.85 | 3.64 | 1.12 | 8.61 |
| 22mm diameter, plugged and screwed | m | 0.24 | 4.20 | 3.84 | 1.21 | 9.25 |
| extra over for | | | | | | |
| straight coupling | nr | 0.20 | 3.50 | 1.03 | 0.68 | 5.21 |
| reduced coupling, 22 × 8mm | nr | 0.20 | 3.50 | 6.80 | 1.55 | 11.85 |
| reduced coupling, 22 × 10mm | nr | 0.20 | 3.50 | 4.26 | 1.16 | 8.92 |
| reduced coupling, 22 × 15mm | nr | 0.20 | 3.50 | 1.76 | 0.79 | 6.05 |
| straight female connector, 22mm × 3/4in | nr | 0.20 | 3.50 | 8.06 | 1.73 | 13.29 |
| female reducing connector, 22mm × 1/2in | nr | 0.20 | 3.50 | 12.80 | 2.45 | 18.75 |
| straight male connector, 22mm × 3/4in | nr | 0.20 | 3.50 | 8.45 | 1.79 | 13.74 |
| male reducing connector, 22mm × 1/2in | nr | 0.20 | 3.50 | 13.24 | 2.51 | 19.25 |
| tank connector, 22mm × 1in | nr | 0.20 | 3.50 | 21.81 | 3.80 | 29.11 |
| reducer, 22 × 12mm | nr | 0.20 | 3.50 | 6.80 | 1.55 | 11.85 |
| female adaptor, 22 × 3/4in | nr | 0.20 | 3.50 | 14.59 | 2.71 | 20.80 |
| male adaptor, 22 × 3/4in | nr | 0.20 | 3.50 | 12.49 | 2.40 | 18.39 |
| adaptor, 22 × 3/4in | nr | 0.20 | 3.50 | 20.86 | 3.65 | 28.01 |

|  | Unit | Labour hours | Labour cost £ | Materials £ | O & P £ | Total £ |
|---|---|---|---|---|---|---|
| street elbow, 22mm | nr | 0.20 | 3.50 | 4.05 | 1.13 | 8.68 |
| male elbow, 22mm × 3/4in | nr | 0.20 | 3.50 | 11.83 | 2.30 | 17.63 |
| female elbow, 22mm × 3/4in | nr | 0.20 | 3.50 | 13.68 | 2.58 | 19.76 |
| reducing female elbow, 22mm × 1/2in | nr | 0.20 | 3.50 | 13.68 | 2.58 | 19.76 |
| backplate elbow, 22mm × 3/4in | nr | 0.20 | 3.50 | 21.31 | 3.72 | 28.53 |
| flanged bend, 22mm × 3/4in | nr | 0.20 | 3.50 | 24.50 | 4.20 | 32.20 |
| overflow bend, 22mm × 3/4in | nr | 0.20 | 3.50 | 30.04 | 5.03 | 38.57 |
| slow bend | nr | 0.20 | 3.50 | 3.15 | 1.00 | 7.65 |
| return bend | nr | 0.20 | 3.50 | 29.23 | 4.91 | 37.64 |
| obtuse elbow | nr | 0.20 | 3.50 | 3.97 | 1.12 | 8.59 |
| tee, reduced branch largest end 22mm | nr | 0.24 | 4.20 | 15.16 | 2.90 | 22.26 |
| tee, both ends reduced, largest end 22mm | nr | 0.24 | 4.20 | 13.09 | 2.59 | 19.88 |
| cross tee, 22mm | nr | 0.24 | 4.20 | 28.05 | 4.84 | 37.09 |
| sweep tee, 90 degrees, 22mm | nr | 0.24 | 4.20 | 21.58 | 3.87 | 29.65 |
| offset tee, 22mm | nr | 0.24 | 4.20 | 26.68 | 4.63 | 35.51 |
| double sweep tee, 22mm | nr | 0.24 | 4.20 | 25.84 | 4.51 | 34.55 |
| equal tee, 22mm | nr | 0.24 | 4.20 | 10.05 | 2.14 | 16.39 |
| stop end, 22mm | nr | 0.20 | 3.50 | 4.56 | 1.21 | 9.27 |
| tap connector, 22mm × 1/2in | nr | 0.20 | 3.50 | 18.98 | 3.37 | 25.85 |
| bent tap connector, 22mm × 1/2in | nr | 0.20 | 3.50 | 23.12 | 3.99 | 30.61 |
| bent tap connector, 22mm × 3/4in | nr | 0.20 | 3.50 | 11.56 | 2.26 | 17.32 |
| bent union adaptor, 22mm × 1in | nr | 0.20 | 3.50 | 15.44 | 2.84 | 21.78 |
| bent male union connector, 22mm × 3/4in | nr | 0.20 | 3.50 | 14.02 | 2.63 | 20.15 |
| bent female union connector, 22mm × 3/4in | nr | 0.20 | 3.50 | 25.16 | 4.30 | 32.96 |
| made bend | nr | 0.20 | 3.50 | 0.00 | 0.53 | 4.03 |

|  | Unit | Labour hours | Labour cost £ | Materials £ | O & P £ | Total £ |
|---|---|---|---|---|---|---|
| **Stop valves** | | | | | | |
| Gunmetal stop valve with brass headwork, copper × copper | | | | | | |
| 15mm | nr | 0.30 | 5.25 | 8.75 | 2.10 | 16.10 |
| 22mm | nr | 0.35 | 6.13 | 16.35 | 3.37 | 25.85 |
| 28mm | nr | 0.40 | 7.00 | 46.48 | 8.02 | 61.50 |
| Gunmetal lockshield stop valve with brass headwork, copper × copper | | | | | | |
| 22mm | nr | 0.35 | 6.13 | 33.10 | 5.88 | 45.11 |
| 28mm | nr | 0.40 | 7.00 | 51.47 | 8.77 | 67.24 |
| **Copper pipework, compression joints** | | | | | | |
| Copper pipe, compression joints and fittings, clips at 1250mm centres | | | | | | |
| 15mm diameter, to timber | m | 0.20 | 3.50 | 1.91 | 0.81 | 6.22 |
| 15mm diameter, plugged and screwed | m | 0.22 | 3.85 | 2.11 | 0.89 | 6.85 |
| extra over for | | | | | | |
| bent union radiator, 15mm × 1/2in | nr | 0.18 | 3.15 | 11.41 | 2.18 | 16.74 |
| air-release elbow, 15mm | nr | 0.18 | 3.15 | 14.70 | 2.68 | 20.53 |
| slow bend, 15mm | nr | 0.18 | 3.15 | 14.35 | 2.63 | 20.13 |
| male elbow, 15mm × 1/2in | nr | 0.18 | 3.15 | 3.99 | 1.07 | 8.21 |
| male elbow, 15mm × 3/4in | nr | 0.18 | 3.15 | 12.19 | 2.30 | 17.64 |
| male elbow, BSP parallel thread, 15mm × 1/2in | nr | 0.18 | 3.15 | 3.99 | 1.07 | 8.21 |
| female elbow, 15mm × 1/2in | nr | 0.18 | 3.15 | 6.12 | 1.39 | 10.66 |
| offset tee, 15 × 15 × 15mm | nr | 0.18 | 3.15 | 26.45 | 4.44 | 34.04 |
| male adaptor, 15mm × 1/2in | nr | 0.18 | 3.15 | 4.83 | 1.20 | 9.18 |
| reducing set, 15 × 8mm | nr | 0.18 | 3.15 | 3.10 | 0.94 | 7.19 |
| reducing set, 15 × 10mm | nr | 0.18 | 3.15 | 3.10 | 0.94 | 7.19 |
| reducing set, 15 × 12mm | nr | 0.18 | 3.15 | 3.45 | 0.99 | 7.59 |

| | Unit | Labour hours | Labour cost £ | Materials £ | O & P £ | Total £ |
|---|---|---|---|---|---|---|
| 22mm diameter, to timber | m | 0.21 | 3.68 | 3.91 | 1.14 | 8.72 |
| 22mm diameter, plugged and screwed | m | 0.23 | 4.03 | 4.11 | 1.22 | 9.36 |
| extra over for | | | | | | |
| air-release elbow, 22mm | nr | 0.18 | 3.15 | 19.40 | 3.38 | 25.93 |
| slow bend, 22mm | nr | 0.18 | 3.15 | 23.12 | 3.94 | 30.21 |
| male elbow, 22mm × 3/4in | nr | 0.18 | 3.15 | 5.16 | 1.25 | 9.56 |
| male elbow, 22mm × 1in | nr | 0.18 | 3.15 | 10.08 | 1.98 | 15.21 |
| male elbow, BSP parallel thread, 22mm × 3/4in | nr | 0.18 | 3.15 | 5.16 | 1.25 | 9.56 |
| male elbow, BSP parallel thread, 22mm × 1in | nr | 0.18 | 3.15 | 10.08 | 1.98 | 15.21 |
| female elbow, 22mm × 3/4in | nr | 0.18 | 3.15 | 8.85 | 1.80 | 13.80 |
| female elbow, 22mm × 1in | nr | 0.18 | 3.15 | 13.78 | 2.54 | 19.47 |
| bent cylinder connector, 22mm × 1in | nr | 0.18 | 3.15 | 21.96 | 3.77 | 28.88 |
| tee with reduced branch, 22 × 22 × 15mm | nr | 0.22 | 3.85 | 12.75 | 2.49 | 19.09 |
| tee, end reduced 22 × 15 × 22mm | nr | 0.22 | 3.85 | 12.75 | 2.49 | 19.09 |
| tee, end and branch reduced, 22 × 15 × 15mm | nr | 0.22 | 3.85 | 14.34 | 2.73 | 20.92 |
| straight swivel connector, 22mm × 1/2in | nr | 0.18 | 3.15 | 11.98 | 2.27 | 17.40 |
| bent swivel connector, 22mm × 1/2in | nr | 0.18 | 3.15 | 15.80 | 2.84 | 21.79 |
| offset tee, 22 × 22 × 15mm | nr | 0.22 | 3.85 | 38.91 | 6.41 | 49.17 |

**Cold water storage tanks**

Galvanised steel cistern with cover, reference

| | Unit | Labour hours | Labour cost £ | Materials £ | O & P £ | Total £ |
|---|---|---|---|---|---|---|
| SC10, 18 litres | nr | 0.60 | 10.50 | 59.28 | 10.47 | 80.25 |
| SC15, 36 litres | nr | 0.65 | 11.38 | 64.21 | 11.34 | 86.92 |
| SC20, 54 litres | nr | 0.65 | 11.38 | 71.02 | 12.36 | 94.75 |
| SC25, 68 litres | nr | 0.70 | 12.25 | 79.98 | 13.83 | 106.06 |
| SC30, 86 litres | nr | 0.75 | 13.13 | 86.40 | 14.93 | 114.45 |
| SC40, 114 litres | nr | 0.80 | 14.00 | 90.08 | 15.61 | 119.69 |
| SC50, 154 litres | nr | 0.85 | 14.88 | 110.08 | 18.74 | 143.70 |

| | Unit | Labour hours | Labour cost £ | Materials £ | O & P £ | Total £ |
|---|---|---|---|---|---|---|
| Cutting holes for connectors | | | | | | |
| 13mm | nr | 0.10 | 1.75 | 0.00 | 0.26 | 2.01 |
| 18mm | nr | 0.12 | 2.10 | 0.00 | 0.32 | 2.42 |
| 25mm | nr | 0.15 | 2.63 | 0.00 | 0.39 | 3.02 |
| 32mm | nr | 0.20 | 3.50 | 0.00 | 0.53 | 4.03 |
| Plastic cistern with lid, reference | | | | | | |
| PC4, 18 litres | nr | 0.60 | 10.50 | 59.28 | 10.47 | 80.25 |
| PC15, 68 litres | nr | 0.70 | 12.25 | 79.98 | 13.83 | 106.06 |
| PC20, 91 litres | nr | 0.75 | 13.13 | 84.42 | 14.63 | 112.18 |
| PC25, 114 litres | nr | 0.80 | 14.00 | 90.08 | 15.61 | 119.69 |
| PC40, 182 litres | nr | 0.90 | 15.75 | 110.02 | 18.87 | 144.64 |
| PC50, 227 litres | nr | 1.00 | 17.50 | 120.22 | 20.66 | 158.38 |

**Hot water copper cylinders**

| | Unit | Labour hours | Labour cost £ | Materials £ | O & P £ | Total £ |
|---|---|---|---|---|---|---|
| Direct cylinders, insulated, reference | | | | | | |
| ref 5, 98 litres | nr | 0.50 | 8.75 | 128.46 | 20.58 | 157.79 |
| ref 7, 120 litres | nr | 0.55 | 9.63 | 134.97 | 21.69 | 166.28 |
| ref 8, 148 litres | nr | 0.65 | 11.38 | 164.21 | 26.34 | 201.92 |
| ref 9, 166 litres | nr | 0.85 | 14.88 | 180.02 | 29.23 | 224.13 |
| Indirect cylinders, insulated, reference | | | | | | |
| ref 2, 96 litres | nr | 0.50 | 8.75 | 131.31 | 21.01 | 161.07 |
| ref 3, 114 litres | nr | 0.55 | 9.63 | 142.70 | 22.85 | 175.17 |
| ref 7, 117 litres | nr | 0.55 | 9.63 | 149.97 | 23.94 | 183.53 |
| ref 8, 140 litres | nr | 0.65 | 11.38 | 169.97 | 27.20 | 208.55 |
| ref 9, 162 litres | nr | 0.70 | 12.25 | 198.21 | 31.57 | 242.03 |

|  | Unit | Labour hours | Labour cost £ | Materials £ | O & P £ | Total £ |
|---|---|---|---|---|---|---|
| **Insulation** | | | | | | |
| Preformed pipe lagging, fire-retardant foam 13mm thick for pipe size | | | | | | |
| 15mm | m | 0.08 | 1.40 | 0.50 | 0.29 | 2.19 |
| 22mm | m | 0.09 | 1.58 | 0.66 | 0.34 | 2.57 |
| 28mm | m | 0.10 | 1.75 | 0.72 | 0.37 | 2.84 |
| Preformed pipe lagging, fire-retardant foam 25mm thick for pipe size | | | | | | |
| 15mm | m | 0.10 | 1.75 | 1.80 | 0.53 | 4.08 |
| 22mm | m | 0.11 | 1.93 | 2.16 | 0.61 | 4.70 |
| 28mm | m | 0.12 | 2.10 | 2.80 | 0.74 | 5.64 |
| 50mm glass-fibre filled insulating jacket, fixing bands to tank size | | | | | | |
| 445 × 305 × 300mm | nr | 0.35 | 6.13 | 7.20 | 2.00 | 15.32 |
| 495 × 368 × 362mm | nr | 0.40 | 7.00 | 8.91 | 2.39 | 18.30 |
| 630 × 450 × 420mm | nr | 0.45 | 7.88 | 9.16 | 2.56 | 19.59 |
| 665 × 490 × 515mm | nr | 0.50 | 8.75 | 11.02 | 2.97 | 22.74 |
| 995 × 605 × 595mm | nr | 0.70 | 12.25 | 13.00 | 3.79 | 29.04 |
| **Boilers** | | | | | | |
| Gas-fired wall-mounted boilers for domestic central heating and indirect hot water | | | | | | |
| Balanced flue, output | | | | | | |
| 30,000 BTU | nr | 4.80 | 84.00 | 485.07 | 85.36 | 654.43 |
| 40,000 BTU | nr | 4.80 | 84.00 | 559.01 | 96.45 | 739.46 |
| 50,000 BTU | nr | 4.80 | 84.00 | 636.85 | 108.13 | 828.98 |
| 60,000 BTU | nr | 4.80 | 84.00 | 738.15 | 123.32 | 945.47 |
| Fan flue, output | | | | | | |
| 30,000 BTU | nr | 4.20 | 73.50 | 517.59 | 88.66 | 679.75 |
| 40,000 BTU | nr | 4.20 | 73.50 | 582.91 | 98.46 | 754.87 |
| 50,000 BTU | nr | 4.20 | 73.50 | 636.03 | 106.43 | 815.96 |
| 60,000 BTU | nr | 4.20 | 73.50 | 688.39 | 114.28 | 876.17 |
| 70,000 BTU | nr | 4.20 | 73.50 | 800.49 | 131.10 | 1005.09 |
| 80,000 BTU | nr | 4.20 | 73.50 | 1230.31 | 195.57 | 1499.38 |

| | Unit | Labour hours | Labour cost £ | Materials £ | O & P £ | Total £ |
|---|---|---|---|---|---|---|
| **Gas-fired floor-standing boilers for domestic central heating and indirect hot water** | | | | | | |
| Fanned balanced flue, output | | | | | | |
| 40,000 BTU | nr | 4.80 | 84.00 | 867.00 | 142.65 | 1093.65 |
| 50,000 BTU | nr | 4.80 | 84.00 | 675.85 | 113.98 | 873.83 |
| 60,000 BTU | nr | 4.80 | 84.00 | 681.49 | 114.82 | 880.31 |
| 70,000 BTU | nr | 4.80 | 84.00 | 717.32 | 120.20 | 921.52 |
| 80,000 BTU | nr | 4.80 | 84.00 | 985.56 | 160.43 | 1229.99 |
| Fanned chimney flue, output | | | | | | |
| 40,000 BTU | nr | 4.20 | 73.50 | 501.42 | 86.24 | 661.16 |
| 50,000 BTU | nr | 4.20 | 73.50 | 547.67 | 93.18 | 714.35 |
| 60,000 BTU | nr | 4.20 | 73.50 | 604.18 | 101.65 | 779.33 |
| 70,000 BTU | nr | 4.20 | 73.50 | 787.96 | 129.22 | 990.68 |
| 80,000 BTU | nr | 4.20 | 73.50 | 1004.70 | 161.73 | 1239.93 |
| **Gas-fired wall-mounted combination boiler for domestic central heating and hot water** | | | | | | |
| Fan flue, output | | | | | | |
| 30,400-81,900 BTU | nr | 4.80 | 84.00 | 746.79 | 124.62 | 955.41 |
| 35,500-95,000 BTU | nr | 4.80 | 84.00 | 801.70 | 132.86 | 1018.56 |
| Condensing fan flue | | | | | | |
| 25,200-92,800 BTU | nr | 4.80 | 84.00 | 938.77 | 153.42 | 1176.19 |
| 25,900-93,800 BTU | nr | 4.80 | 84.00 | 1057.05 | 171.16 | 1312.21 |
| Conventional flue | | | | | | |
| 31,000-82,000 BTU | nr | 4.80 | 84.00 | 767.28 | 127.69 | 978.97 |
| **Gas-fired floor-standing combination boiler for domestic central heating and hot water** | | | | | | |
| Fan flue, output | | | | | | |
| 37,500-81,900 BTU | nr | 4.80 | 84.00 | 1236.36 | 198.05 | 1518.41 |

| | Unit | Labour hours | Labour cost £ | Materials £ | O & P £ | Total £ |
|---|---|---|---|---|---|---|
| **Oil-fired boiler for domestic central heating and hot water** | | | | | | |
| | | | | | | |
| Balanced flue, output | | | | | | |
| 40,000-48,000 BTU | nr | 4.80 | 84.00 | 867.79 | 142.77 | 1094.56 |
| 50,000-65,000 BTU | nr | 4.80 | 84.00 | 970.74 | 158.21 | 1212.95 |
| 75,000-85,000 BTU | nr | 4.80 | 84.00 | 1084.44 | 175.27 | 1343.71 |
| 88,000-110,000 BTU | nr | 4.80 | 84.00 | 1215.45 | 194.92 | 1494.37 |
| | | | | | | |
| Conventional flue, output | | | | | | |
| 52,000 BTU | nr | 4.80 | 84.00 | 866.95 | 142.64 | 1093.59 |
| 70,000 BTU | nr | 4.80 | 84.00 | 970.21 | 158.13 | 1212.34 |
| 90,000 BTU | nr | 4.80 | 84.00 | 1084.22 | 175.23 | 1343.45 |
| 120,000 BTU | nr | 4.80 | 84.00 | 1215.45 | 194.92 | 1494.37 |
| **Oil-fired combination boiler for domestic central heating and hot water** | | | | | | |
| | | | | | | |
| 50,000-65,000 BTU | nr | 4.80 | 84.00 | 1386.56 | 220.58 | 1691.14 |
| 75,000-85,000 BTU | nr | 4.80 | 84.00 | 1506.83 | 238.62 | 1829.45 |
| 88,000-110,000 BTU | nr | 4.80 | 84.00 | 1667.22 | 262.68 | 2013.90 |
| | | | | | | |
| Conventional flue, output | | | | | | |
| 52,000 BTU | nr | 4.80 | 84.00 | 1310.33 | 209.15 | 1603.48 |
| 70,000 BTU | nr | 4.80 | 84.00 | 1386.48 | 220.57 | 1691.05 |
| 90,000 BTU | nr | 4.80 | 84.00 | 1505.87 | 238.48 | 1828.35 |
| 120,000 BTU | nr | 4.80 | 84.00 | 1667.22 | 262.68 | 2013.90 |
| **Oil storage tanks** | | | | | | |
| | | | | | | |
| Standard domestic pattern plastic oil storage tank | | | | | | |
| 1380 × 1780 × 890mm | | | | | | |
| capacity 1200 litres | nr | 1.50 | 26.25 | 195.76 | 33.30 | 255.31 |
| 745 × 1700 × 1380mm | | | | | | |
| capacity 1225 litres | nr | 1.50 | 26.25 | 278.44 | 45.70 | 350.39 |
| 1190mm diameter × 1400mm | | | | | | |
| high, capacity 1300 litres | nr | 1.50 | 26.25 | 196.65 | 33.44 | 256.34 |

| | Unit | Labour hours | Labour cost £ | Materials £ | O & P £ | Total £ |
|---|---|---|---|---|---|---|
| 1130 × 2055 × 1145mm capacity 1800 litres | nr | 1.75 | 30.63 | 344.68 | 56.30 | 431.60 |
| 1265 × 2250 × 1320mm capacity 2500 litres | nr | 2.00 | 35.00 | 429.17 | 69.63 | 533.80 |
| 1640mm diameter × 1750mm high, capacity 3600 litres | nr | 2.50 | 43.75 | 473.27 | 77.55 | 594.57 |
| 2480mm diameter × 2640 high, capacity 10,000 litres | nr | 3.00 | 52.50 | 1371.42 | 213.59 | 1637.51 |

Standard domestic pattern plastic bunded oil storage tank

| | Unit | Labour hours | Labour cost £ | Materials £ | O & P £ | Total £ |
|---|---|---|---|---|---|---|
| 1380 × 1780 × 890mm capacity capacity1200 litres | nr | 1.50 | 26.25 | 262.45 | 43.31 | 332.01 |
| 745 × 1700 × 1380mm capacity 1225 litres | nr | 1.50 | 26.25 | 344.51 | 55.61 | 426.37 |
| 1190mm diameter × 1400 high, capacity 1300 litres | nr | 1.50 | 26.25 | 804.57 | 124.62 | 955.44 |
| 1130 × 2055 × 1145mm capacity 1800 litres | nr | 1.75 | 30.63 | 840.85 | 130.72 | 1002.20 |
| 1265 × 2250 × 1320mm capacity 2500 litres | nr | 2.00 | 35.00 | 933.80 | 145.32 | 1114.12 |

## Radiators

Single convector radiator fixed to concealed brackets plugged and screwed to brickwork

300mm high, length

| | Unit | Labour hours | Labour cost £ | Materials £ | O & P £ | Total £ |
|---|---|---|---|---|---|---|
| 400mm | nr | 0.90 | 15.75 | 17.60 | 5.00 | 38.35 |
| 800mm | nr | 1.05 | 18.38 | 29.23 | 7.14 | 54.75 |
| 1200mm | nr | 1.15 | 20.13 | 42.05 | 9.33 | 71.50 |
| 1600mm | nr | 1.30 | 22.75 | 54.87 | 11.64 | 89.26 |
| 2000mm | nr | 1.40 | 24.50 | 66.51 | 13.65 | 104.66 |

450mm high, length

| | Unit | Labour hours | Labour cost £ | Materials £ | O & P £ | Total £ |
|---|---|---|---|---|---|---|
| 500mm | nr | 1.00 | 17.50 | 17.22 | 5.21 | 39.93 |
| 1000mm | nr | 1.15 | 20.13 | 30.06 | 7.53 | 57.71 |
| 1400mm | nr | 1.25 | 21.88 | 40.12 | 9.30 | 71.29 |
| 1800mm | nr | 1.40 | 24.50 | 57.92 | 12.36 | 94.78 |
| 2000mm | nr | 1.50 | 26.25 | 77.17 | 15.51 | 118.93 |

| | Unit | Labour hours | Labour cost £ | Materials £ | O & P £ | Total £ |
|---|---|---|---|---|---|---|
| **600mm high, length** | | | | | | |
| 500mm | nr | 1.10 | 19.25 | 21.24 | 6.07 | 46.56 |
| 1000mm | nr | 1.25 | 21.88 | 36.52 | 8.76 | 67.15 |
| 1400mm | nr | 1.35 | 23.63 | 49.69 | 11.00 | 84.31 |
| 1800mm | nr | 1.50 | 26.25 | 72.33 | 14.79 | 113.37 |
| 2000mm | nr | 1.60 | 28.00 | 96.12 | 18.62 | 142.74 |
| **700mm high, length** | | | | | | |
| 500mm | nr | 1.20 | 21.00 | 23.63 | 6.69 | 51.32 |
| 1000mm | nr | 1.35 | 23.63 | 41.29 | 9.74 | 74.65 |
| 1400mm | nr | 1.45 | 25.38 | 75.94 | 15.20 | 116.51 |
| 1800mm | nr | 1.60 | 28.00 | 98.55 | 18.98 | 145.53 |
| 2000mm | nr | 1.70 | 29.75 | 109.25 | 20.85 | 159.85 |

**Double convector radiator fixed to concealed brackets plugged and screwed to brickwork**

| | Unit | Labour hours | Labour cost £ | Materials £ | O & P £ | Total £ |
|---|---|---|---|---|---|---|
| **300mm high, length** | | | | | | |
| 400mm | nr | 0.90 | 15.75 | 29.54 | 6.79 | 52.08 |
| 800mm | nr | 1.05 | 18.38 | 51.90 | 10.54 | 80.82 |
| 1200mm | nr | 1.15 | 20.13 | 74.28 | 14.16 | 108.57 |
| 1600mm | nr | 1.30 | 22.75 | 100.22 | 18.45 | 141.42 |
| 2000mm | nr | 1.40 | 24.50 | 123.79 | 22.24 | 170.53 |
| **450mm high, length** | | | | | | |
| 500mm | nr | 1.00 | 17.50 | 27.96 | 6.82 | 52.28 |
| 1000mm | nr | 1.15 | 20.13 | 51.55 | 10.75 | 82.43 |
| 1400mm | nr | 1.25 | 21.88 | 79.51 | 15.21 | 116.59 |
| 1800mm | nr | 1.40 | 24.50 | 128.34 | 22.93 | 175.77 |
| 2000mm | nr | 1.50 | 26.25 | 141.49 | 25.16 | 192.90 |
| **600mm high, length** | | | | | | |
| 500mm | nr | 1.10 | 19.25 | 33.53 | 7.92 | 60.70 |
| 1000mm | nr | 1.25 | 21.88 | 63.48 | 12.80 | 98.16 |
| 1400mm | nr | 1.35 | 23.63 | 96.67 | 18.04 | 138.34 |
| 1800mm | nr | 1.50 | 26.25 | 158.35 | 27.69 | 212.29 |
| 2000mm | nr | 1.60 | 28.00 | 176.58 | 30.69 | 235.27 |

| | Unit | Labour hours | Labour cost £ | Materials £ | O & P £ | Total £ |
|---|---|---|---|---|---|---|
| **Double panel radiator fixed to concealed brackets plugged and screwed to brickwork** | | | | | | |
| | | | | | | |
| **300mm high, length** | | | | | | |
| 400mm | nr | 0.90 | 15.75 | 23.56 | 5.90 | 45.21 |
| 800mm | nr | 1.05 | 18.38 | 41.16 | 8.93 | 68.47 |
| 1200mm | nr | 1.15 | 20.13 | 59.96 | 12.01 | 92.10 |
| 1600mm | nr | 1.30 | 22.75 | 78.74 | 15.22 | 116.71 |
| 2000mm | nr | 1.40 | 24.50 | 96.35 | 18.13 | 138.98 |
| | | | | | | |
| **450mm high, length** | | | | | | |
| 500mm | nr | 1.00 | 17.50 | 23.19 | 6.10 | 46.79 |
| 1000mm | nr | 1.15 | 20.13 | 40.80 | 9.14 | 70.06 |
| 1400mm | nr | 1.25 | 21.88 | 62.80 | 12.70 | 97.38 |
| 1800mm | nr | 1.40 | 24.50 | 100.88 | 18.81 | 144.19 |
| 2000mm | nr | 1.50 | 26.25 | 111.66 | 20.69 | 158.60 |
| | | | | | | |
| **600mm high, length** | | | | | | |
| 500mm | nr | 1.10 | 19.25 | 27.57 | 7.02 | 53.84 |
| 1000mm | nr | 1.25 | 21.88 | 50.35 | 10.83 | 83.06 |
| 1400mm | nr | 1.35 | 23.63 | 78.39 | 15.30 | 117.32 |
| 1800mm | nr | 1.50 | 26.25 | 125.64 | 22.78 | 174.67 |
| 2000mm | nr | 1.60 | 28.00 | 139.60 | 25.14 | 192.74 |

| | Unit | Labour hours | Labour cost £ | Materials £ | O & P £ | Total £ |
|---|---|---|---|---|---|---|
| **GLAZING** | | | | | | |
| **Clear float glass** | | | | | | |
| In wood with putty and sprigs | | | | | | |
| under 0.15m2, thickness | | | | | | |
| 3mm | m2 | 0.90 | 14.40 | 37.68 | 7.81 | 59.89 |
| 4mm | m2 | 0.90 | 14.40 | 38.02 | 7.86 | 60.28 |
| 5mm | m2 | 0.90 | 14.40 | 54.21 | 10.29 | 78.90 |
| 6mm | m2 | 1.00 | 16.00 | 56.08 | 10.81 | 82.89 |
| 10mm | m2 | 1.05 | 16.80 | 108.91 | 18.86 | 144.57 |
| over 0.15m2, thickness | | | | | | |
| 3mm | m2 | 0.60 | 9.60 | 37.68 | 7.09 | 54.37 |
| 4mm | m2 | 0.60 | 9.60 | 38.02 | 7.14 | 54.76 |
| 5mm | m2 | 0.60 | 9.60 | 54.21 | 9.57 | 73.38 |
| 6mm | m2 | 0.65 | 10.40 | 56.08 | 9.97 | 76.45 |
| 10mm | m2 | 0.70 | 11.20 | 108.91 | 18.02 | 138.13 |
| In wood with pinned beads | | | | | | |
| under 0.15m2, thickness | | | | | | |
| 3mm | m2 | 1.10 | 17.60 | 38.06 | 8.35 | 64.01 |
| 4mm | m2 | 1.10 | 17.60 | 38.40 | 8.40 | 64.40 |
| 5mm | m2 | 1.10 | 17.60 | 57.75 | 11.30 | 86.65 |
| 6mm | m2 | 1.20 | 19.20 | 58.75 | 11.69 | 89.64 |
| 10mm | m2 | 1.25 | 20.00 | 109.98 | 19.50 | 149.48 |
| over 0.15m2, thickness | | | | | | |
| 3mm | m2 | 0.80 | 12.80 | 38.06 | 7.63 | 58.49 |
| 4mm | m2 | 0.80 | 12.80 | 38.40 | 7.68 | 58.88 |
| 5mm | m2 | 0.80 | 12.80 | 57.75 | 10.58 | 81.13 |
| 6mm | m2 | 0.90 | 14.40 | 58.75 | 10.97 | 84.12 |
| 10mm | m2 | 0.95 | 15.20 | 109.98 | 18.78 | 143.96 |

|  | Unit | Labour hours | Labour cost £ | Materials £ | O & P £ | Total £ |
|---|---|---|---|---|---|---|
| **In wood with screwed beads** | | | | | | |
| under 0.15m2, thickness | | | | | | |
| 3mm | m2 | 1.30 | 20.80 | 38.06 | 8.83 | 67.69 |
| 4mm | m2 | 1.30 | 20.80 | 38.40 | 8.88 | 68.08 |
| 5mm | m2 | 1.30 | 20.80 | 57.75 | 11.78 | 90.33 |
| 6mm | m2 | 1.40 | 22.40 | 58.75 | 12.17 | 93.32 |
| 10mm | m2 | 1.45 | 23.20 | 109.98 | 19.98 | 153.16 |
| over 0.15m2, thickness | | | | | | |
| 3mm | m2 | 1.00 | 16.00 | 38.06 | 8.11 | 62.17 |
| 4mm | m2 | 1.00 | 16.00 | 38.40 | 8.16 | 62.56 |
| 5mm | m2 | 1.00 | 16.00 | 57.75 | 11.06 | 84.81 |
| 6mm | m2 | 1.05 | 16.80 | 58.75 | 11.33 | 86.88 |
| 10mm | m2 | 1.10 | 17.60 | 109.98 | 19.14 | 146.72 |
| **In metal with putty** | | | | | | |
| under 0.15m2, thickness | | | | | | |
| 3mm | m2 | 0.90 | 14.40 | 37.68 | 7.81 | 59.89 |
| 4mm | m2 | 0.90 | 14.40 | 38.02 | 7.86 | 60.28 |
| 5mm | m2 | 0.90 | 14.40 | 54.21 | 10.29 | 78.90 |
| 6mm | m2 | 1.00 | 16.00 | 56.08 | 10.81 | 82.89 |
| 10mm | m2 | 1.05 | 16.80 | 108.91 | 18.86 | 144.57 |
| over 0.15m2, thickness | | | | | | |
| 3mm | m2 | 0.60 | 9.60 | 37.68 | 7.09 | 54.37 |
| 4mm | m2 | 0.60 | 9.60 | 38.02 | 7.14 | 54.76 |
| 5mm | m2 | 0.60 | 9.60 | 54.21 | 9.57 | 73.38 |
| 6mm | m2 | 0.65 | 10.40 | 56.08 | 9.97 | 76.45 |
| 10mm | m2 | 0.70 | 11.20 | 108.91 | 18.02 | 138.13 |
| **In metal with clips and putty** | | | | | | |
| under 0.15m2, thickness | | | | | | |
| 3mm | m2 | 0.95 | 15.20 | 37.68 | 7.93 | 60.81 |
| 4mm | m2 | 0.95 | 15.20 | 38.02 | 7.98 | 61.20 |
| 5mm | m2 | 0.95 | 15.20 | 54.21 | 10.41 | 79.82 |
| 6mm | m2 | 1.05 | 16.80 | 56.08 | 10.93 | 83.81 |
| 10mm | m2 | 1.10 | 17.60 | 108.91 | 18.98 | 145.49 |

|  | Unit | Labour hours | Labour cost £ | Materials £ | O & P £ | Total £ |
|---|---|---|---|---|---|---|
| over 0.15m2, thickness |  |  |  |  |  |  |
| 3mm | m2 | 0.65 | 10.40 | 37.68 | 7.21 | 55.29 |
| 4mm | m2 | 0.65 | 10.40 | 38.02 | 7.26 | 55.68 |
| 5mm | m2 | 0.65 | 10.40 | 54.21 | 9.69 | 74.30 |
| 6mm | m2 | 0.70 | 11.20 | 56.08 | 10.09 | 77.37 |
| 10mm | m2 | 0.75 | 12.00 | 108.91 | 18.14 | 139.05 |
| **In metal with screwed metal beads** |  |  |  |  |  |  |
| under 0.15m2, thickness |  |  |  |  |  |  |
| 3mm | m2 | 1.30 | 20.80 | 38.06 | 8.83 | 67.69 |
| 4mm | m2 | 1.30 | 20.80 | 38.40 | 8.88 | 68.08 |
| 5mm | m2 | 1.30 | 20.80 | 57.75 | 11.78 | 90.33 |
| 6mm | m2 | 1.40 | 22.40 | 58.75 | 12.17 | 93.32 |
| 10mm | m2 | 1.45 | 23.20 | 109.98 | 19.98 | 153.16 |
| over 0.15m2, thickness |  |  |  |  |  |  |
| 3mm | m2 | 1.00 | 16.00 | 38.06 | 8.11 | 62.17 |
| 4mm | m2 | 1.00 | 16.00 | 38.40 | 8.16 | 62.56 |
| 5mm | m2 | 1.00 | 16.00 | 57.75 | 11.06 | 84.81 |
| 6mm | m2 | 1.05 | 16.80 | 58.75 | 11.33 | 86.88 |
| 10mm | m2 | 1.10 | 17.60 | 109.98 | 19.14 | 146.72 |

## White patterned glass

In wood with putty and sprigs

| | Unit | Labour hours | Labour cost £ | Materials £ | O & P £ | Total £ |
|---|---|---|---|---|---|---|
| under 0.15m2, thickness |  |  |  |  |  |  |
| 4mm | m2 | 0.90 | 14.40 | 37.20 | 7.74 | 59.34 |
| 6mm | m2 | 1.00 | 16.00 | 56.16 | 10.82 | 82.98 |
| over 0.15m2, thickness |  |  |  |  |  |  |
| 4mm | m2 | 0.60 | 9.60 | 37.20 | 7.02 | 53.82 |
| 6mm | m2 | 0.65 | 10.40 | 56.16 | 9.98 | 76.54 |
| In wood with pinned beads |  |  |  |  |  |  |
| under 0.15m2, thickness |  |  |  |  |  |  |
| 4mm | m2 | 1.10 | 17.60 | 39.43 | 8.55 | 65.58 |
| 6mm | m2 | 1.20 | 19.20 | 59.53 | 11.81 | 90.54 |

| | Unit | Labour hours | Labour cost £ | Materials £ | O & P £ | Total £ |
|---|---|---|---|---|---|---|
| over 0.15m2, thickness | | | | | | |
| 4mm | m2 | 0.80 | 12.80 | 39.43 | 7.83 | 60.06 |
| 6mm | m2 | 0.85 | 13.60 | 59.53 | 10.97 | 84.10 |
| **In wood with screwed beads** | | | | | | |
| under 0.15m2, thickness | | | | | | |
| 4mm | m2 | 1.30 | 20.80 | 39.43 | 9.03 | 69.26 |
| 6mm | m2 | 1.40 | 22.40 | 59.53 | 12.29 | 94.22 |
| over 0.15m2, thickness | | | | | | |
| 4mm | m2 | 1.00 | 16.00 | 39.43 | 8.31 | 63.74 |
| 6mm | m2 | 1.50 | 24.00 | 59.53 | 12.53 | 96.06 |
| **In metal with putty** | | | | | | |
| under 0.15m2, thickness | | | | | | |
| 4mm | m2 | 0.90 | 14.40 | 37.20 | 7.74 | 59.34 |
| 6mm | m2 | 1.00 | 16.00 | 56.16 | 10.82 | 82.98 |
| over 0.15m2, thickness | | | | | | |
| 4mm | m2 | 0.60 | 9.60 | 37.20 | 7.02 | 53.82 |
| 6mm | m2 | 0.65 | 10.40 | 56.16 | 9.98 | 76.54 |
| **In metal with clips and putty** | | | | | | |
| under 0.15m2, thickness | | | | | | |
| 4mm | m2 | 0.95 | 15.20 | 37.20 | 7.86 | 60.26 |
| 6mm | m2 | 1.05 | 16.80 | 56.16 | 10.94 | 83.90 |
| over 0.15m2, thickness | | | | | | |
| 4mm | m2 | 0.65 | 10.40 | 37.20 | 7.14 | 54.74 |
| 6mm | m2 | 0.70 | 11.20 | 56.16 | 10.10 | 77.46 |
| **In metal with screwed metal beads** | | | | | | |
| under 0.15m2, thickness | | | | | | |
| 4mm | m2 | 1.30 | 20.80 | 39.43 | 9.03 | 69.26 |
| 6mm | m2 | 1.40 | 22.40 | 59.53 | 12.29 | 94.22 |

| | Unit | Labour hours | Labour cost £ | Materials £ | O & P £ | Total £ |
|---|---|---|---|---|---|---|
| over 0.15m2, thickness | | | | | | |
| 4mm | m2 | 1.00 | 16.00 | 39.43 | 8.31 | 63.74 |
| 6mm | m2 | 1.05 | 16.80 | 59.53 | 11.45 | 87.78 |

## Georgian wired cast glass

In wood with putty and sprigs

| | Unit | Labour hours | Labour cost £ | Materials £ | O & P £ | Total £ |
|---|---|---|---|---|---|---|
| under 0.15m2, thickness | | | | | | |
| 7mm | m2 | 0.90 | 14.40 | 46.17 | 9.09 | 69.66 |
| over 0.15m2, thickness | | | 0.00 | | | |
| 7mm | m2 | 0.65 | 10.40 | 46.17 | 8.49 | 65.06 |

In wood with pinned beads

| | Unit | Labour hours | Labour cost £ | Materials £ | O & P £ | Total £ |
|---|---|---|---|---|---|---|
| under 0.15m2, thickness | | | | | | |
| 7mm | m2 | 1.20 | 19.20 | 48.94 | 10.22 | 78.36 |
| over 0.15m2, thickness | | | | | | |
| 7mm | m2 | 0.90 | 14.40 | 48.94 | 9.50 | 72.84 |

In wood with screwed beads

| | Unit | Labour hours | Labour cost £ | Materials £ | O & P £ | Total £ |
|---|---|---|---|---|---|---|
| under 0.15m2, thickness | | | | | | |
| 7mm | m2 | 1.40 | 22.40 | 48.94 | 10.70 | 82.04 |
| over 0.15m2, thickness | | | | | | |
| 7mm | m2 | 1.05 | 16.80 | 48.94 | 9.86 | 75.60 |

In metal with putty and clips

| | Unit | Labour hours | Labour cost £ | Materials £ | O & P £ | Total £ |
|---|---|---|---|---|---|---|
| under 0.15m2, thickness | | | | | | |
| 7mm | m2 | 1.05 | 16.80 | 46.17 | 9.45 | 72.42 |
| over 0.15m2, thickness | | | | | | |
| 7mm | m2 | 0.70 | 11.20 | 46.17 | 8.61 | 65.98 |

| | Unit | Labour hours | Labour cost £ | Materials £ | O & P £ | Total £ |
|---|---|---|---|---|---|---|
| **In metal with screwed metal beads** | | | | | | |
| under 0.15m2, thickness 7mm | m2 | 1.40 | 22.40 | 48.94 | 10.70 | 82.04 |
| over 0.15m2, thickness 7mm | m2 | 1.05 | 16.80 | 48.94 | 9.86 | 75.60 |
| **Georgian wired polished glass** | | | | | | |
| In wood with putty and sprigs | | | | | | |
| under 0.15m2, thickness 6mm | m2 | 0.90 | 14.40 | 93.04 | 16.12 | 123.56 |
| over 0.15m2, thickness 6mm | m2 | 0.65 | 10.40 | 93.04 | 15.52 | 118.96 |
| In wood with pinned beads | | | | | | |
| under 0.15m2, thickness 6mm | m2 | 1.20 | 19.20 | 95.17 | 17.16 | 131.53 |
| over 0.15m2, thickness 6mm | m2 | 0.90 | 14.40 | 95.17 | 16.44 | 126.01 |
| In wood with screwed beads | | | | | | |
| under 0.15m2, thickness 6mm | m2 | 1.40 | 22.40 | 95.17 | 17.64 | 135.21 |
| over 0.15m2, thickness 6mm | m2 | 1.05 | 16.80 | 95.17 | 16.80 | 128.77 |

| | Unit | Labour hours | Labour cost £ | Materials £ | O & P £ | Total £ |
|---|---|---|---|---|---|---|
| **In metal with putty and clips** | | | | | | |
| under 0.15m2, thickness | | | | | | |
| 7mm | m2 | 1.05 | 16.80 | 93.04 | 16.48 | 126.32 |
| over 0.15m2, thickness | | | | | | |
| 7mm | m2 | 0.70 | 11.20 | 93.04 | 15.64 | 119.88 |
| **In metal with screwed metal beads** | | | | | | |
| under 0.15m2, thickness | | | | | | |
| 7mm | m2 | 1.40 | 22.40 | 95.17 | 17.64 | 135.21 |
| over 0.15m2, thickness | | | | | | |
| 7mm | m2 | 1.05 | 16.80 | 95.17 | 16.80 | 128.77 |
| **Clear laminated safety glass** | | | | | | |
| In wood with pinned beads | | | | | | |
| under 0.15m2, thickness | | | | | | |
| 4.4mm | m2 | 1.10 | 17.60 | 73.14 | 13.61 | 104.35 |
| 6.4mm | m2 | 1.20 | 19.20 | 89.41 | 16.29 | 124.90 |
| 8.8mm | m2 | 1.30 | 20.80 | 112.82 | 20.04 | 153.66 |
| over 0.15m2, thickness | | | | | | |
| 4.4mm | m2 | 0.80 | 12.80 | 73.14 | 12.89 | 98.83 |
| 6.4mm | m2 | 0.90 | 14.40 | 89.41 | 15.57 | 119.38 |
| 8.8mm | m2 | 1.00 | 16.00 | 112.82 | 19.32 | 148.14 |
| In wood with screwed beads | | | | | | |
| under 0.15m2, thickness | | | | | | |
| 4.4mm | m2 | 1.30 | 18.20 | 77.53 | 12.26 | 77.53 |
| 6.4mm | m2 | 1.40 | 19.60 | 97.77 | 17.61 | 134.98 |
| 8.8mm | m2 | 1.50 | 24.00 | 119.59 | 21.54 | 165.13 |

| | Unit | Labour hours | Labour cost £ | Materials £ | O & P £ | Total £ |
|---|---|---|---|---|---|---|
| over 0.15m2, thickness | | | | | | |
| 4mm | m2 | 1.00 | 16.00 | 77.53 | 14.03 | 107.56 |
| 4.4mm | m2 | 1.00 | 16.00 | 97.77 | 17.07 | 130.84 |
| 6.4mm | m2 | 1.05 | 16.80 | 119.59 | 20.46 | 156.85 |
| In metal with screwed beads | | | | | | |
| under 0.15m2, thickness | | | | | | |
| 4mm | m2 | 1.30 | 20.80 | 77.53 | 14.75 | 113.08 |
| 4.4mm | m2 | 1.30 | 20.80 | 97.77 | 17.79 | 136.36 |
| 6.4mm | m2 | 1.40 | 22.40 | 119.59 | 21.30 | 163.29 |
| over 0.15m2, thickness | | | | | | |
| 4mm | m2 | 1.00 | 16.00 | 77.53 | 14.03 | 107.56 |
| 4.4mm | m2 | 1.00 | 16.00 | 97.77 | 17.07 | 130.84 |
| 6.4mm | m2 | 1.05 | 16.80 | 119.59 | 20.46 | 156.85 |

|  | Unit | Labour hours | Labour cost £ | Materials £ | O & P £ | Total £ |
|---|---|---|---|---|---|---|
| **PAINTING AND WALLPAPERING** | | | | | | |
| **On internal surfaces before fixing** | | | | | | |
| One coat of primer on wood surfaces | | | | | | |
| exceeding 300mm girth | m2 | 0.16 | 2.56 | 0.60 | 0.47 | 3.63 |
| not exceeding 150mm girth | m | 0.04 | 0.64 | 0.10 | 0.11 | 0.85 |
| 150-300mm girth | m | 0.06 | 0.96 | 0.20 | 0.17 | 1.33 |
| isolated areas not exceeding 0.5m2 | nr | 0.10 | 1.60 | 0.30 | 0.29 | 2.19 |
| One coat of primer on metal surfaces | | | | | | |
| exceeding 300mm girth | m2 | 0.16 | 2.56 | 0.78 | 0.50 | 3.84 |
| not exceeding 150mm girth | m | 0.04 | 0.64 | 0.13 | 0.12 | 0.89 |
| 150-300mm girth | m | 0.06 | 0.96 | 0.26 | 0.18 | 1.40 |
| isolated areas not exceeding 0.5m2 | nr | 0.10 | 1.60 | 0.39 | 0.30 | 2.29 |
| One coat of wood preservative on wood surfaces | | | | | | |
| exceeding 300mm girth | m2 | 0.14 | 2.24 | 0.40 | 0.40 | 3.04 |
| not exceeding 150mm girth | m | 0.03 | 0.48 | 0.07 | 0.08 | 0.63 |
| 150-300mm girth | m | 0.05 | 0.80 | 0.13 | 0.14 | 1.07 |
| isolated areas not exceeding 0.5m2 | nr | 0.08 | 1.28 | 0.20 | 0.22 | 1.70 |
| **On internal walls and ceilings** | | | | | | |
| One coat of emulsion paint on | | | | | | |
| brickwork or blockwork walls | m2 | 0.20 | 3.20 | 0.69 | 0.58 | 4.47 |
| brickwork or blockwork walls in staircase areas | m2 | 0.26 | 4.16 | 0.69 | 0.73 | 5.58 |
| plastered walls | m2 | 0.12 | 1.92 | 0.46 | 0.36 | 2.74 |
| plastered walls in staircase | m2 | 0.16 | 2.56 | 0.46 | 0.45 | 3.47 |

| | Unit | Labour hours | Labour cost £ | Materials £ | O & P £ | Total £ |
|---|---|---|---|---|---|---|
| plastered ceilings | m2 | 0.20 | 3.20 | 0.46 | 0.55 | 4.21 |
| plastered ceilings in staircase areas | m2 | 0.24 | 3.84 | 0.46 | 0.65 | 4.95 |
| embossed papered walls | m2 | 0.16 | 2.56 | 0.57 | 0.47 | 3.60 |
| embossed papered walls in staircase areas | m2 | 0.20 | 3.20 | 0.57 | 0.57 | 4.34 |
| **Two coats of emulsion paint on** | | | | | | |
| brickwork or blockwork wall | m2 | 0.28 | 4.48 | 1.39 | 0.88 | 6.75 |
| brickwork or blockwork walls in staircase areas | m2 | 0.32 | 5.12 | 1.39 | 0.98 | 7.49 |
| plastered walls | m2 | 0.16 | 2.56 | 0.96 | 0.53 | 4.05 |
| plastered walls in staircase | m2 | 0.20 | 3.20 | 0.96 | 0.62 | 4.78 |
| plastered ceilings | m2 | 0.24 | 3.84 | 0.96 | 0.72 | 5.52 |
| plastered ceilings in staircase areas | m2 | 0.28 | 4.48 | 0.96 | 0.82 | 6.26 |
| embossed papered walls | m2 | 0.24 | 3.84 | 1.16 | 0.75 | 5.75 |
| embossed papered walls in staircase areas | m2 | 0.26 | 4.16 | 1.16 | 0.80 | 6.12 |
| **One mist coat and two coats of emulsion paint on** | | | | | | |
| brickwork or blockwork walls | m2 | 0.36 | 5.76 | 1.74 | 1.13 | 8.63 |
| brickwork or blockwork walls in staircase areas | m2 | 0.40 | 6.40 | 1.74 | 1.22 | 9.36 |
| plastered walls | m2 | 0.24 | 3.84 | 1.16 | 0.75 | 5.75 |
| plastered walls in staircase areas | m2 | 0.28 | 4.48 | 1.16 | 0.85 | 6.49 |
| plastered ceilings | m2 | 0.32 | 5.12 | 1.16 | 0.94 | 7.22 |
| plastered ceilings in staircase areas | m2 | 0.36 | 5.76 | 1.16 | 1.04 | 7.96 |
| embossed papered walls | m2 | 0.30 | 4.80 | 1.30 | 0.92 | 7.02 |
| embossed papered walls in staircase areas | m2 | 0.34 | 5.44 | 1.30 | 1.01 | 7.75 |

| | Unit | Labour hours | Labour cost £ | Materials £ | O & P £ | Total £ |
|---|---|---|---|---|---|---|
| One coat sealer primer, one undercoat and one coat gloss on | | | | | | |
| plastered walls | m2 | 0.42 | 6.72 | 1.52 | 1.24 | 9.48 |
| plastered walls in staircase | m2 | 0.46 | 7.36 | 1.52 | 1.33 | 10.21 |
| plastered ceilings | m2 | 0.48 | 7.68 | 1.52 | 1.38 | 10.58 |
| plastered ceilings in staircase areas | m2 | 0.52 | 8.32 | 1.52 | 1.48 | 11.32 |

**On internal woodwork**

One coat primer on wood general surfaces

| | Unit | Labour hours | Labour cost £ | Materials £ | O & P £ | Total £ |
|---|---|---|---|---|---|---|
| exceeding 300mm girth | m2 | 0.18 | 2.88 | 0.57 | 0.52 | 3.97 |
| not exceeding 150mm girth | m | 0.04 | 0.64 | 0.10 | 0.11 | 0.85 |
| 150-300mm girth | m | 0.06 | 0.96 | 0.18 | 0.17 | 1.31 |
| isolated areas not exceeding 0.5m2 | nr | 0.10 | 1.60 | 0.29 | 0.28 | 2.17 |

One coat primer on wood windows, glazed doors and the like

| | Unit | Labour hours | Labour cost £ | Materials £ | O & P £ | Total £ |
|---|---|---|---|---|---|---|
| panes area not exceeding 0.1m2 | m2 | 0.35 | 5.60 | 0.23 | 0.87 | 6.70 |
| panes area 0.1 to 0.5m2 | m2 | 0.30 | 4.80 | 0.18 | 0.75 | 5.73 |
| panes area 0.5 to 1m2 | m2 | 0.25 | 4.00 | 0.16 | 0.62 | 4.78 |
| panes area exceeding 1m2 | m2 | 0.22 | 3.52 | 0.14 | 0.55 | 4.21 |

One coat primer on wood frames, linings, skirtings and the like

| | Unit | Labour hours | Labour cost £ | Materials £ | O & P £ | Total £ |
|---|---|---|---|---|---|---|
| exceeding 300mm girth | m2 | 0.20 | 3.20 | 0.53 | 0.56 | 4.29 |
| not exceeding 150mm girth | m | 0.05 | 0.80 | 0.08 | 0.13 | 1.01 |
| 150-300mm girth | m | 0.08 | 1.28 | 0.16 | 0.22 | 1.66 |
| isolated areas not exceeding 0.5m2 | nr | 0.11 | 1.76 | 0.23 | 0.30 | 2.29 |

| | Unit | Labour hours | Labour cost £ | Materials £ | O & P £ | Total £ |
|---|---|---|---|---|---|---|
| **One coat undercoat on primed wood general surfaces** | | | | | | |
| exceeding 300mm girth | m2 | 0.18 | 2.88 | 0.57 | 0.52 | 3.97 |
| not exceeding 150mm girth | m | 0.04 | 0.64 | 0.10 | 0.11 | 0.85 |
| 150-300mm girth | m | 0.06 | 0.96 | 0.18 | 0.17 | 1.31 |
| isolated areas not exceeding 0.5m2 | nr | 0.10 | 1.60 | 0.29 | 0.28 | 2.17 |
| **One coat undercoat on primed wood windows, glazed doors and the like** | | | | | | |
| panes area not exceeding 0.1m2 | m2 | 0.35 | 5.60 | 0.23 | 0.87 | 6.70 |
| panes area 0.1 to 0.5m2 | m2 | 0.30 | 4.80 | 0.21 | 0.75 | 5.76 |
| panes area 0.5 to 1m2 | m2 | 0.25 | 4.00 | 0.18 | 0.63 | 4.81 |
| panes area exceeding 1m2 | m2 | 0.22 | 3.52 | 0.17 | 0.55 | 4.24 |
| **One coat undercoat on primed wood frames, linings, skirtings and the like** | | | | | | |
| exceeding 300mm girth | m2 | 0.20 | 3.20 | 0.57 | 0.57 | 4.34 |
| not exceeding 150mm girth | m | 0.05 | 0.80 | 0.10 | 0.14 | 1.04 |
| 150-300mm girth | m | 0.08 | 1.28 | 0.18 | 0.22 | 1.68 |
| isolated areas not exceeding 0.5m2 | nr | 0.11 | 1.76 | 0.28 | 0.31 | 2.35 |
| **Two coats undercoat on primed wood general surfaces** | | | | | | |
| exceeding 300mm girth | m2 | 0.36 | 5.76 | 1.28 | 1.06 | 8.10 |
| not exceeding 150mm girth | m | 0.08 | 1.28 | 0.21 | 0.22 | 1.71 |
| 150-300mm girth | m | 0.12 | 1.92 | 0.42 | 0.35 | 2.69 |
| isolated areas not exceeding 0.5m2 | nr | 0.20 | 0.00 3.20 | 0.64 | 0.58 | 4.42 |

| | Unit | Labour hours | Labour cost £ | Materials £ | O & P £ | Total £ |
|---|---|---|---|---|---|---|
| **Two coats undercoat on primed wood windows, glazed doors and the like** | | | | | | |
| panes area not exceeding 0.1m2 | m2 | 0.60 | 9.60 | 0.50 | 1.52 | 11.62 |
| panes area 0.1 to 0.5m2 | m2 | 0.50 | 8.00 | 0.42 | 1.26 | 9.68 |
| panes area 0.5 to 1m2 | m2 | 0.40 | 6.40 | 0.46 | 1.03 | 7.89 |
| panes area exceeding 1m2 | m2 | 0.30 | 4.80 | 0.36 | 0.77 | 5.93 |
| **Two coats undercoat on primed wood frames, linings, skirtings and the like** | | | | | | |
| exceeding 300mm girth | m2 | 0.40 | 6.40 | 1.28 | 1.15 | 8.83 |
| not exceeding 150mm girth | m | 0.10 | 1.60 | 0.21 | 0.27 | 2.08 |
| 150-300mm girth | m | 0.16 | 2.56 | 0.42 | 0.45 | 3.43 |
| isolated areas not exceeding 0.5m2 | nr | 0.22 | 3.54 | 0.64 | 0.63 | 4.80 |
| **One coat gloss on undercoated wood general surfaces** | | | | | | |
| exceeding 300mm girth | m2 | 0.18 | 2.88 | 0.64 | 0.53 | 4.05 |
| not exceeding 150mm girth | m | 0.04 | 0.64 | 0.11 | 0.11 | 0.86 |
| 150-300mm girth | m | 0.06 | 0.96 | 0.21 | 0.18 | 1.35 |
| isolated areas not exceeding 0.5m2 | nr | 0.10 | 1.60 | 0.32 | 0.29 | 2.21 |
| **One coat gloss on undercoated wood windows, glazed doors and the like** | | | | | | |
| panes area not exceeding 0.1m2 | m2 | 0.35 | 5.60 | 0.26 | 0.88 | 6.74 |
| panes area 0.1 to 0.5m2 | m2 | 0.30 | 4.80 | 0.24 | 0.76 | 5.80 |
| panes area 0.5 to 1m2 | m2 | 0.25 | 4.00 | 0.20 | 0.63 | 4.83 |
| panes area exceeding 1m2 | m2 | 0.22 | 3.52 | 0.19 | 0.56 | 4.27 |

| | Unit | Labour hours | Labour cost £ | Materials £ | O & P £ | Total £ |
|---|---|---|---|---|---|---|
| **One coat gloss on undercoated wood frames, linings, skirtings and the like** | | | | | | |
| exceeding 300mm girth | m2 | 0.20 | 3.20 | 0.64 | 0.58 | 4.42 |
| not exceeding 150mm girth | m | 0.05 | 0.80 | 0.11 | 0.14 | 1.05 |
| 150-300mm girth | m | 0.08 | 1.28 | 0.21 | 0.22 | 1.71 |
| isolated areas not exceeding 0.5m2 | nr | 0.11 | 1.76 | 0.32 | 0.31 | 2.39 |
| **One coat primer, one coat undercoat and one coat gloss on wood general surfaces** | | | | | | |
| exceeding 300mm girth | m2 | 0.54 | 8.64 | 1.94 | 1.59 | 12.17 |
| not exceeding 150mm girth | m | 0.12 | 1.92 | 0.32 | 0.34 | 2.58 |
| 150-300mm girth | m | 0.18 | 2.88 | 0.63 | 0.53 | 4.04 |
| isolated areas not exceeding 0.5m2 | nr | 0.30 | 4.80 | 0.47 | 0.79 | 6.06 |
| **One coat primer, one coat undercoat and one coat gloss on wood windows, glazed doors and the like** | | | | | | |
| panes area not exceeding 0.1m2 | m2 | 0.75 | 12.00 | 0.74 | 1.91 | 14.65 |
| panes area 0.1 to 0.5m2 | m2 | 0.60 | 9.60 | 0.62 | 1.53 | 11.75 |
| panes area 0.5 to 1m2 | m2 | 0.50 | 8.00 | 0.56 | 1.28 | 9.84 |
| panes area exceeding 1m2 | m2 | 0.40 | 6.40 | 0.48 | 1.03 | 7.91 |
| **One coat primer, one coat undercoat and one coat gloss on wood frames, linings, skirtings and the like** | | | | | | |
| exceeding 300mm girth | m2 | 0.60 | 9.60 | 1.90 | 1.73 | 13.23 |
| not exceeding 150mm girth | m | 0.15 | 2.40 | 0.30 | 0.41 | 3.11 |
| 150-300mm girth | m | 0.24 | 3.84 | 0.60 | 0.67 | 5.11 |
| isolated areas not exceeding 0.5m2 | nr | 0.33 | 5.28 | 0.95 | 0.93 | 7.16 |

| | Unit | Labour hours | Labour cost £ | Materials £ | O & P £ | Total £ |
|---|---|---|---|---|---|---|
| **One coat primer, two coats undercoat and one coat gloss on wood general surfaces** | | | | | | |
| exceeding 300mm girth | m2 | 0.72 | 11.52 | 2.46 | 2.10 | 16.08 |
| not exceeding 150mm girth | m | 0.16 | 2.56 | 0.41 | 0.45 | 3.42 |
| 150-300mm girth | m | 0.24 | 3.84 | 0.82 | 0.70 | 5.36 |
| isolated areas not exceeding 0.5m2 | nr | 0.40 | 6.40 | 1.43 | 1.17 | 9.00 |
| **One coat primer, two coats undercoat and one coat gloss on wood windows, glazed doors and the like** | | | | | | |
| panes area not exceeding 0.1m2 | m2 | 1.00 | 16.00 | 0.98 | 2.55 | 19.53 |
| panes area 0.1 to 0.5m2 | m2 | 0.80 | 12.80 | 0.90 | 2.06 | 15.76 |
| panes area 0.5 to 1m2 | m2 | 0.70 | 11.20 | 0.78 | 1.80 | 13.78 |
| panes area exceeding 1m2 | m2 | 0.60 | 9.60 | 0.70 | 1.55 | 11.85 |
| **One coat primer, two coats undercoat and one coat gloss on wood frames, linings, skirtings and the like** | | | | | | |
| exceeding 300mm girth | m2 | 0.80 | 12.80 | 2.40 | 2.28 | 17.48 |
| not exceeding 150mm girth | m | 0.20 | 3.20 | 0.40 | 0.54 | 4.14 |
| 150-300mm girth | m | 0.32 | 5.12 | 0.80 | 0.89 | 6.81 |
| isolated areas not exceeding 0.5m2 | nr | 0.44 | 7.04 | 1.20 | 1.24 | 9.48 |

| | Unit | Labour hours | Labour cost £ | Materials £ | O & P £ | Total £ |
|---|---|---|---|---|---|---|
| **On internal metalwork** | | | | | | |
| One coat primer on metal general surfaces | | | | | | |
| exceeding 300mm girth | m2 | 0.18 | 2.88 | 0.70 | 0.54 | 4.12 |
| not exceeding 150mm girth | m | 0.04 | 0.64 | 0.12 | 0.11 | 0.87 |
| 150-300mm girth | m | 0.06 | 0.96 | 0.23 | 0.18 | 1.37 |
| isolated areas not exceeding 0.5m2 | nr | 0.10 | 1.60 | 0.35 | 0.29 | 2.24 |
| One coat primer on metal windows, glazed doors and the like | | | | | | |
| panes area not exceeding 0.1m2 | m2 | 0.35 | 5.60 | 0.26 | 0.88 | 6.74 |
| panes area 0.1 to 0.5m2 | m2 | 0.30 | 4.80 | 0.24 | 0.76 | 5.80 |
| panes area 0.5 to 1m2 | m2 | 0.25 | 4.00 | 0.20 | 0.63 | 4.83 |
| panes area exceeding 1m2 | m2 | 0.22 | 3.52 | 0.19 | 0.56 | 4.27 |
| One coat primer on structural steelwork | | | | | | |
| exceeding 300mm girth | m2 | 0.24 | 3.84 | 0.70 | 0.68 | 5.22 |
| One coat primer on metal radiators | | | | | | |
| exceeding 300mm girth | m2 | 0.18 | 2.88 | 0.70 | 0.54 | 4.12 |
| One coat undercoat on metal general surfaces | | | | | | |
| exceeding 300mm girth | m2 | 0.18 | 2.88 | 0.64 | 0.53 | 4.05 |
| not exceeding 150mm girth | m | 0.04 | 0.64 | 0.11 | 0.11 | 0.86 |
| 150-300mm girth | m | 0.06 | 0.96 | 0.21 | 0.18 | 1.35 |
| isolated areas not exceeding 0.5m2 | nr | 0.10 | 1.60 | 0.32 | 0.29 | 2.21 |

|  | Unit | Labour hours | Labour cost £ | Materials £ | O & P £ | Total £ |
|---|---|---|---|---|---|---|
| **One coat undercoat on metal windows, glazed doors and the like** | | | | | | |
| panes area not exceeding 0.1m2 | m2 | 0.35 | 5.60 | 0.24 | 0.88 | 6.72 |
| panes area 0.1 to 0.5m2 | m2 | 0.30 | 4.80 | 0.20 | 0.75 | 5.75 |
| panes area 0.5 to 1m2 | m2 | 0.25 | 4.00 | 0.18 | 0.63 | 4.81 |
| panes area exceeding 1m2 | m2 | 0.22 | 3.52 | 0.16 | 0.55 | 4.23 |
| **One coat undercoat on structural steelwork** | | | | | | |
| exceeding 300mm girth | m2 | 0.24 | 3.84 | 0.62 | 0.67 | 5.13 |
| **One coat undercoat on metal radiators** | | | | | | |
| exceeding 300mm girth | m2 | 0.18 | 2.88 | 0.62 | 0.53 | 4.03 |
| **Two coats undercoat on metal general surfaces** | | | | | | |
| exceeding 300mm girth | m2 | 0.18 | 2.88 | 1.20 | 0.61 | 4.69 |
| not exceeding 150mm girth | m | 0.04 | 0.64 | 0.20 | 0.13 | 0.97 |
| 150-300mm girth | m | 0.06 | 0.96 | 0.40 | 0.20 | 1.56 |
| isolated areas not exceeding 0.5m2 | nr | 0.10 | 1.60 | 0.60 | 0.33 | 2.53 |
| **Two coats undercoat on metal windows, glazed doors and the like** | | | | | | |
| panes area not exceeding 0.1m2 | m2 | 0.35 | 5.60 | 0.48 | 0.91 | 6.99 |
| panes area 0.1 to 0.5m2 | m2 | 0.30 | 4.80 | 0.40 | 0.78 | 5.98 |
| panes area 0.5 to 1m2 | m2 | 0.25 | 4.00 | 0.32 | 0.65 | 4.97 |
| panes area exceeding 1m2 | m2 | 0.22 | 3.52 | 0.24 | 0.56 | 4.32 |

| | Unit | Labour hours | Labour cost £ | Materials £ | O & P £ | Total £ |
|---|---|---|---|---|---|---|
| **Two coats undercoat on structural steelwork** | | | | | | |
| exceeding 300mm girth | m2 | 0.24 | 3.84 | 1.20 | 0.76 | 5.80 |
| **Two coats undercoat on metal radiators** | | | | | | |
| exceeding 300mm girth | m2 | 0.18 | 2.88 | 1.20 | 0.61 | 4.69 |
| **One coat gloss on metal general surfaces** | | | | | | |
| exceeding 300mm girth | m2 | 0.18 | 2.88 | 0.32 | 0.48 | 3.68 |
| not exceeding 150mm girth | m | 0.04 | 0.64 | 0.05 | 0.10 | 0.79 |
| 150-300mm girth | m | 0.06 | 0.96 | 0.10 | 0.16 | 1.22 |
| isolated areas not exceeding 0.5m2 | nr | 0.10 | 1.60 | 0.16 | 0.26 | 2.02 |
| **One coat gloss on metal windows, glazed doors and the like** | | | | | | |
| panes area not exceeding 0.1m2 | m2 | 0.35 | 5.60 | 0.24 | 0.88 | 6.72 |
| panes area 0.1 to 0.5m2 | m2 | 0.30 | 4.80 | 0.20 | 0.75 | 5.75 |
| panes area 0.5 to 1m2 | m2 | 0.25 | 4.00 | 0.18 | 0.63 | 4.81 |
| panes area exceeding 1m2 | m2 | 0.22 | 3.52 | 0.16 | 0.55 | 4.23 |
| **One coat gloss on structural steelwork** | | | | | | |
| exceeding 300mm girth | m2 | 0.24 | 3.84 | 0.64 | 0.67 | 5.15 |
| **One coat gloss on metal radiators** | | | | | | |
| exceeding 300mm girth | m2 | 0.18 | 2.88 | 0.64 | 0.53 | 4.05 |

| | Unit | Labour hours | Labour cost £ | Materials £ | O & P £ | Total £ |
|---|---|---|---|---|---|---|
| **One coat primer, one coat undercoat and one coat gloss on metal general surfaces** | | | | | | |
| exceeding 300mm girth | m2 | 0.54 | 8.64 | 1.96 | 1.59 | 12.19 |
| not exceeding 150mm girth | m | 0.12 | 1.92 | 0.34 | 0.34 | 2.60 |
| 150-300mm girth | m | 0.18 | 2.88 | 0.68 | 0.53 | 4.09 |
| isolated areas not exceeding 0.5m2 | nr | 0.30 | 4.80 | 0.98 | 0.87 | 6.65 |
| **One coat primer, one coat undercoat and one coat gloss on wood windows, glazed doors and the like** | | | | | | |
| panes not exceeding 0.1m2 | m2 | 0.75 | 12.00 | 0.80 | 1.92 | 14.72 |
| panes area 0.1 to 0.5m2 | m2 | 0.60 | 9.60 | 0.13 | 1.46 | 11.19 |
| panes area 0.5 to 1m2 | m2 | 0.50 | 8.00 | 0.26 | 1.24 | 9.50 |
| panes area exceeding 1m2 | m2 | 0.40 | 6.40 | 0.40 | 1.02 | 7.82 |
| **One coat primer, one coat undercoat and one coat gloss on structural steelwork** | | | | | | |
| exceeding 300mm girth | m2 | 0.60 | 9.60 | 1.96 | 1.73 | 13.29 |
| **One coat primer, one coat undercoat and one coat gloss on metal radiators** | | | | | | |
| exceeding 300mm girth | m2 | 0.52 | 8.32 | 1.96 | 1.54 | 11.82 |
| **One coat primer, two coats undercoat and one coat gloss on metal general surfaces** | | | | | | |
| exceeding 300mm girth | m2 | 0.54 | 8.64 | 2.50 | 1.67 | 12.81 |
| not exceeding 150mm girth | m | 0.12 | 1.92 | 0.42 | 0.35 | 2.69 |
| 150-300mm girth | m | 0.18 | 2.88 | 0.84 | 0.56 | 4.28 |
| isolated areas | nr | 0.30 | 4.80 | 1.25 | 0.91 | 6.96 |

| | Unit | Labour hours | Labour cost £ | Materials £ | O & P £ | Total £ |
|---|---|---|---|---|---|---|
| **One coat primer, two coats undercoat and one coat gloss on metal windows, glazed doors and the like** | | | | | | |
| panes area not exceeding | | | | | | |
| 0.1m2 | m2 | 0.75 | 12.00 | 0.98 | 1.95 | 14.93 |
| panes area 0.1 to 0.5m2 | m2 | 0.60 | 9.60 | 0.90 | 1.58 | 12.08 |
| panes area 0.5 to 1m2 | m2 | 0.50 | 8.00 | 0.82 | 1.32 | 10.14 |
| panes area exceeding 1m2 | m2 | 0.40 | 6.40 | 0.74 | 1.07 | 8.21 |
| **One coat primer, two coats undercoat and one coat gloss on metal radiators** | | | | | | |
| exceeding 300mm girth | m2 | 0.72 | 11.52 | 2.46 | 2.10 | 16.08 |

**On external woodwork**

| | Unit | Labour hours | Labour cost £ | Materials £ | O & P £ | Total £ |
|---|---|---|---|---|---|---|
| **One coat primer on wood general surfaces** | | | | | | |
| exceeding 300mm girth | m2 | 0.24 | 3.84 | 0.64 | 0.67 | 5.15 |
| not exceeding 150mm girth | m | 0.06 | 0.96 | 0.11 | 0.16 | 1.23 |
| 150-300mm girth | m | 0.08 | 1.28 | 0.21 | 0.22 | 1.71 |
| isolated areas not exceeding | | | | | | |
| 0.5m2 | nr | 0.02 | 0.32 | 0.32 | 0.10 | 0.74 |
| **One coat primer on wood windows, glazed doors and the like** | | | | | | |
| panes area not exceeding | | | | | | |
| 0.1m2 | m2 | 0.40 | 6.40 | 0.24 | 1.00 | 7.64 |
| panes area 0.1 to 0.5m2 | m2 | 0.35 | 5.60 | 0.20 | 0.87 | 6.67 |
| panes area 0.5 to 1m2 | m2 | 0.30 | 4.80 | 0.18 | 0.75 | 5.73 |
| panes area exceeding 1m2 | m2 | 0.25 | 4.00 | 0.16 | 0.62 | 4.78 |

| | Unit | Labour hours | Labour cost £ | Materials £ | O & P £ | Total £ |
|---|---|---|---|---|---|---|
| **One coat primer on wood frames and the like** | | | | | | |
| exceeding 300mm girth | m2 | 0.26 | 4.16 | 0.64 | 0.72 | 5.52 |
| not exceeding 150mm girth | m | 0.07 | 1.12 | 0.11 | 0.18 | 1.41 |
| 150-300mm girth | m | 0.10 | 1.60 | 0.22 | 0.27 | 2.09 |
| isolated areas not exceeding 0.5m2 | nr | 0.14 | 2.24 | 0.32 | 0.38 | 2.94 |
| **One coat undercoat on primed wood general surfaces** | | | | | | |
| exceeding 300mm girth | m2 | 0.24 | 3.84 | 0.64 | 0.67 | 5.15 |
| not exceeding 150mm girth | m | 0.06 | 0.96 | 0.11 | 0.16 | 1.23 |
| 150-300mm girth | m | 0.08 | 1.28 | 0.21 | 0.22 | 1.71 |
| isolated areas not exceeding 0.5m2 | nr | 0.12 | 1.92 | 0.32 | 0.34 | 2.58 |
| **One coat undercoat on primed wood windows, glazed doors and the like** | | | | | | |
| panes area not exceeding 0.1m2 | m2 | 0.40 | 6.40 | 0.26 | 1.00 | 7.66 |
| panes area 0.1 to 0.5m2 | m2 | 0.35 | 5.60 | 0.24 | 0.88 | 6.72 |
| panes area 0.5 to 1m2 | m2 | 0.30 | 4.80 | 0.20 | 0.75 | 5.75 |
| panes area exceeding 1m2 | m2 | 0.25 | 4.00 | 0.19 | 0.63 | 4.82 |
| **One coat undercoat on primed wood frames and the like** | | | | | | |
| exceeding 300mm girth | m2 | 0.26 | 4.16 | 0.64 | 0.72 | 5.52 |
| not exceeding 150mm girth | m | 0.07 | 1.12 | 0.11 | 0.18 | 1.41 |
| 150-300mm girth | m | 0.10 | 1.60 | 0.21 | 0.27 | 2.08 |
| isolated areas not exceeding 0.5m2 | nr | 0.14 | 2.24 | 0.32 | 0.38 | 2.94 |

|  | Unit | Labour hours | Labour cost £ | Materials £ | O & P £ | Total £ |
|---|---|---|---|---|---|---|
| **Two coats undercoat on primed wood general surfaces** | | | | | | |
| exceeding 300mm girth | m2 | 0.48 | 7.68 | 1.20 | 1.33 | 10.21 |
| not exceeding 150mm girth | m | 0.12 | 1.92 | 0.20 | 0.32 | 2.44 |
| 150-300mm girth | m | 0.16 | 2.56 | 0.40 | 0.44 | 3.40 |
| isolated areas not exceeding 0.5m2 | nr | 0.24 | 3.84 | 0.60 | 0.67 | 5.11 |
| **Two coats undercoat on primed wood windows, glazed doors and the like** | | | | | | |
| panes area not exceeding 0.1m2 | m2 | 0.80 | 12.80 | 0.48 | 1.99 | 15.27 |
| panes area 0.1 to 0.5m2 | m2 | 0.70 | 11.20 | 0.46 | 1.75 | 13.41 |
| panes area 0.5 to 1m2 | m2 | 0.60 | 9.60 | 0.44 | 1.51 | 11.55 |
| panes area exceeding 1m2 | m2 | 0.50 | 8.00 | 0.40 | 1.26 | 9.66 |
| **Two coats undercoat on primed wood frames and the like** | | | | | | |
| exceeding 300mm girth | m2 | 0.52 | 8.32 | 1.20 | 1.43 | 10.95 |
| not exceeding 150mm girth | m | 0.14 | 2.24 | 0.20 | 0.37 | 2.81 |
| 150-300mm girth | m | 0.20 | 3.20 | 0.40 | 0.54 | 4.14 |
| isolated areas not exceeding 0.5m2 | nr | 0.32 | 5.12 | 0.60 | 0.86 | 6.58 |
| **One coat gloss on undercoated wood general surfaces** | | | | | | |
| exceeding 300mm girth | m2 | 0.24 | 3.84 | 0.64 | 0.67 | 5.15 |
| not exceeding 150mm girth | m | 0.06 | 0.96 | 0.11 | 0.16 | 1.23 |
| 150-300mm girth | m | 0.08 | 1.28 | 0.21 | 0.22 | 1.71 |
| isolated areas not exceeding 0.5m2 | nr | 0.12 | 1.92 | 0.32 | 0.34 | 2.58 |

|  | Unit | Labour hours | Labour cost £ | Materials £ | O & P £ | Total £ |
|---|---|---|---|---|---|---|
| **One coat gloss on undercoated wood windows, glazed doors and the like** | | | | | | |
| panes area not exceeding 0.1m2 | m2 | 0.40 | 6.40 | 0.24 | 1.00 | 7.64 |
| panes area 0.1 to 0.5m2 | m2 | 0.35 | 5.60 | 0.20 | 0.87 | 6.67 |
| panes area 0.5 to 1m2 | m2 | 0.30 | 4.80 | 0.18 | 0.75 | 5.73 |
| panes area exceeding 1m2 | m2 | 0.25 | 4.00 | 0.16 | 0.62 | 4.78 |
| **One coat gloss on undercoated wood frames and the like** | | | | | | |
| exceeding 300mm girth | m2 | 0.26 | 4.16 | 0.64 | 0.72 | 5.52 |
| not exceeding 150mm girth | m | 0.07 | 1.12 | 0.11 | 0.18 | 1.41 |
| 150-300mm girth | m | 0.10 | 1.60 | 0.21 | 0.27 | 2.08 |
| isolated areas not exceeding 0.5m2 | nr | 0.16 | 2.56 | 0.32 | 0.43 | 3.31 |
| **One coat primer, one coat undercoat and one coat gloss on wood general surfaces** | | | | | | |
| exceeding 300mm girth | m2 | 0.73 | 11.60 | 1.94 | 2.03 | 15.57 |
| not exceeding 150mm girth | m | 0.14 | 2.24 | 0.32 | 0.38 | 2.94 |
| 150-300mm girth | m | 0.24 | 3.84 | 0.63 | 0.67 | 5.14 |
| isolated areas not exceeding 0.5m2 | nr | 0.40 | 6.40 | 0.97 | 1.11 | 8.48 |
| **One coat primer, one coat undercoat and one coat gloss on wood windows, glazed doors and the like** | | | | | | |
| panes area not exceeding 0.1m2 | m2 | 1.20 | 19.20 | 0.74 | 2.99 | 22.93 |
| panes area 0.1 to 0.5m2 | m2 | 1.05 | 16.80 | 0.62 | 2.61 | 20.03 |
| panes area 0.5 to 1m2 | m2 | 0.90 | 14.40 | 0.56 | 2.24 | 17.20 |
| panes area exceeding 1m2 | m2 | 0.75 | 12.00 | 0.48 | 1.87 | 14.35 |

| | Unit | Labour hours | Labour cost £ | Materials £ | O & P £ | Total £ |
|---|---|---|---|---|---|---|
| **One coat primer, one coat undercoat and one coat gloss on wood frames and the like** | | | | | | |
| exceeding 300mm girth | m2 | 0.78 | 12.48 | 1.94 | 2.16 | 16.58 |
| not exceeding 150mm girth | m | 0.21 | 3.36 | 0.32 | 0.55 | 4.23 |
| 150-300mm girth | m | 0.30 | 4.80 | 0.63 | 0.81 | 6.24 |
| isolated areas not exceeding 0.5m2 | nr | 0.48 | 7.68 | 0.97 | 1.30 | 9.95 |
| **One coat primer, two coats undercoat and one coat gloss on wood general surfaces** | | | | | | |
| exceeding 300mm girth | m2 | 0.96 | 15.36 | 2.50 | 2.68 | 20.54 |
| not exceeding 150mm girth | m | 0.25 | 4.00 | 0.42 | 0.66 | 5.08 |
| 150-300mm girth | m | 0.34 | 5.44 | 0.84 | 0.94 | 7.22 |
| isolated areas not exceeding 0.5m2 | nr | 0.48 | 7.68 | 1.25 | 1.34 | 10.27 |
| **One coat primer, two coats undercoat and one coat gloss on wood windows, glazed doors and the like** | | | | | | |
| panes area not exceeding 0.1m2 | m2 | 1.60 | 25.60 | 0.98 | 3.99 | 30.57 |
| panes area 0.1 to 0.5m2 | m2 | 1.40 | 22.40 | 0.90 | 3.50 | 26.80 |
| panes area 0.5 to 1m2 | m2 | 1.20 | 19.20 | 0.82 | 3.00 | 23.02 |
| panes area exceeding 1m2 | m2 | 1.00 | 16.00 | 0.74 | 2.51 | 19.25 |
| **One coat primer, two coats undercoat and one coat gloss on wood frames and the like** | | | | | | |
| exceeding 300mm girth | m2 | 1.04 | 16.64 | 2.50 | 2.87 | 22.01 |
| not exceeding 150mm girth | m | 0.28 | 4.48 | 0.42 | 0.74 | 5.64 |
| 150-300mm girth | m | 0.40 | 6.40 | 0.84 | 1.09 | 8.33 |
| isolated areas | nr | 0.64 | 10.24 | 1.25 | 1.72 | 13.21 |

| | Unit | Labour hours | Labour cost £ | Materials £ | O & P £ | Total £ |
|---|---|---|---|---|---|---|
| **On external metalwork** | | | | | | |
| One coat primer on metal general surfaces | | | | | | |
| exceeding 300mm girth | m2 | 0.24 | 3.84 | 0.70 | 0.68 | 5.22 |
| not exceeding 150mm girth | m | 0.06 | 0.96 | 0.12 | 0.16 | 1.24 |
| 150-300mm girth | m | 0.08 | 1.28 | 0.23 | 0.23 | 1.74 |
| isolated areas not exceeding 0.5m2 | nr | 0.12 | 1.92 | 0.35 | 0.34 | 2.61 |
| One coat primer on metal windows, glazed doors and the like | | | | | | |
| panes area not exceeding 0.1m2 | m2 | 0.40 | 6.40 | 0.26 | 1.00 | 7.66 |
| panes area 0.1 to 0.5m2 | m2 | 0.35 | 5.60 | 0.24 | 0.88 | 6.72 |
| panes area 0.5 to 1m2 | m2 | 0.30 | 4.80 | 0.20 | 0.75 | 5.75 |
| panes area exceeding 1m2 | m2 | 0.25 | 4.00 | 0.14 | 0.62 | 4.76 |
| One coat primer on structural steelwork | | | | | | |
| exceeding 300mm girth | m2 | 0.30 | 4.80 | 0.70 | 0.83 | 6.33 |
| One coat primer on pipes and gutters | | | | | | |
| exceeding 300mm girth | m2 | 0.26 | 4.16 | 0.70 | 0.73 | 5.59 |
| 150-300mm girth | m | 0.08 | 1.28 | 0.12 | 0.21 | 1.61 |
| One coat undercoat on metal general surfaces | | | | | | |
| exceeding 300mm girth | m2 | 0.24 | 3.84 | 0.64 | 0.67 | 5.15 |
| not exceeding 150mm girth | m | 0.06 | 0.96 | 0.11 | 0.16 | 1.23 |
| 150-300mm girth | m | 0.08 | 1.28 | 0.21 | 0.22 | 1.71 |
| isolated areas not exceeding 0.5m2 | nr | 0.12 | 1.92 | 0.32 | 0.34 | 2.58 |

| | Unit | Labour hours | Labour cost £ | Materials £ | O & P £ | Total £ |
|---|---|---|---|---|---|---|
| **One coat undercoat on metal windows, glazed doors and the like** | | | | | | |
| panes area not exceeding 0.1m2 | m2 | 0.40 | 6.40 | 0.24 | 1.00 | 7.64 |
| panes area 0.1 to 0.5m2 | m2 | 0.35 | 5.60 | 0.20 | 0.87 | 6.67 |
| panes area 0.5 to 1m2 | m2 | 0.30 | 4.80 | 0.18 | 0.75 | 5.73 |
| panes area exceeding 1m2 | m2 | 0.25 | 4.00 | 0.16 | 0.62 | 4.78 |
| **One coat undercoat on structural steelwork** | | | | | | |
| exceeding 300mm girth | m2 | 0.30 | 4.80 | 0.62 | 0.81 | 6.23 |
| **One coat primer on pipes and gutters** | | | | | | |
| exceeding 300mm girth | m2 | 0.26 | 4.16 | 0.62 | 0.72 | 5.50 |
| 150-300mm girth | m | 0.08 | 1.28 | 0.10 | 0.21 | 1.59 |
| **Two coats undercoat on metal general surfaces** | | | | | | |
| exceeding 300mm girth | m2 | 0.48 | 7.68 | 1.20 | 1.33 | 10.21 |
| not exceeding 150mm girth | m | 0.12 | 1.92 | 0.20 | 0.32 | 2.44 |
| 150-300mm girth | m | 0.20 | 3.20 | 0.40 | 0.54 | 4.14 |
| isolated areas not exceeding 0.5m2 | nr | 0.24 | 3.84 | 0.60 | 0.67 | 5.11 |
| **Two coats undercoat on metal windows, glazed doors and the like** | | | | | | |
| panes area not exceeding 0.1m2 | m2 | 0.80 | 12.80 | 0.48 | 1.99 | 15.27 |
| panes area 0.1 to 0.5m2 | m2 | 0.70 | 11.20 | 0.40 | 1.74 | 13.34 |
| panes area 0.5 to 1m2 | m2 | 0.60 | 9.60 | 0.32 | 1.49 | 11.41 |
| panes area exceeding 1m2 | m2 | 0.50 | 8.00 | 0.24 | 1.24 | 9.48 |

| | Unit | Labour hours | Labour cost £ | Materials £ | O & P £ | Total £ |
|---|---|---|---|---|---|---|
| **Two coats undercoat on structural steelwork** | | | | | | |
| exceeding 300mm girth | m2 | 0.60 | 9.60 | 1.20 | 1.62 | 12.42 |
| **Two coats undercoat on pipes and gutters** | | | | | | |
| exceeding 300mm girth | m2 | 0.52 | 8.32 | 1.20 | 1.43 | 10.95 |
| 150-300mm girth | m | 0.16 | 2.56 | 0.20 | 0.41 | 3.17 |
| **One coat gloss on metal general surfaces** | | | | | | |
| exceeding 300mm girth | m2 | 0.24 | 3.84 | 0.32 | 0.62 | 4.78 |
| not exceeding 150mm girth | m | 0.06 | 0.96 | 0.05 | 0.15 | 1.16 |
| 150-300mm girth | m | 0.08 | 1.28 | 0.10 | 0.21 | 1.59 |
| isolated areas not exceeding 0.5m2 | nr | 0.12 | 1.92 | 0.16 | 0.31 | 2.39 |
| **One coat gloss on metal windows, glazed doors and the like** | | | | | | |
| panes area not exceeding 0.1m2 | m2 | 0.40 | 6.40 | 0.24 | 1.00 | 7.64 |
| panes area 0.1 to 0.5m2 | m2 | 0.35 | 5.60 | 0.20 | 0.87 | 6.67 |
| panes area 0.5 to 1m2 | m2 | 0.30 | 4.80 | 0.18 | 0.75 | 5.73 |
| panes area exceeding 1m2 | m2 | 0.25 | 4.00 | 0.16 | 0.62 | 4.78 |
| **One coat gloss on structural steelwork** | | | | | | |
| exceeding 300mm girth | m2 | 0.30 | 4.80 | 0.64 | 0.82 | 6.26 |
| **One coat gloss on pipes and gutters** | | | | | | |
| exceeding 300mm girth | m2 | 0.26 | 4.16 | 0.64 | 0.72 | 5.52 |
| 150-300mm girth | m | 0.08 | 1.28 | 0.11 | 0.21 | 1.60 |

| | Unit | Labour hours | Labour cost £ | Materials £ | O & P £ | Total £ |
|---|---|---|---|---|---|---|
| One coat primer, one coat undercoat and one coat gloss on metal general surfaces | | | | | | |
| exceeding 300mm girth | m2 | 0.72 | 11.52 | 1.96 | 2.02 | 15.50 |
| not exceeding 150mm girth | m | 0.18 | 2.88 | 0.34 | 0.48 | 3.70 |
| 150-300mm girth | m | 0.24 | 3.84 | 0.68 | 0.68 | 5.20 |
| isolated areas not exceeding 0.5m2 | nr | 0.36 | 5.76 | 0.98 | 1.01 | 7.75 |
| One coat primer, one coat undercoat and one coat gloss on metal windows, glazed doors and the like | | | | | | |
| panes area not exceeding 0.1m2 | m2 | 1.20 | 19.20 | 0.80 | 3.00 | 23.00 |
| panes area 0.1 to 0.5m2 | m2 | 1.05 | 16.80 | 0.13 | 2.54 | 19.47 |
| panes area 0.5 to 1m2 | m2 | 0.90 | 14.40 | 0.26 | 2.20 | 16.86 |
| panes area exceeding 1m2 | m2 | 0.75 | 12.00 | 0.40 | 1.86 | 14.26 |
| One coat primer, one coat undercoat and one coat gloss on structural steelwork | | | | | | |
| exceeding 300mm girth | m2 | 0.90 | 14.40 | 1.96 | 2.45 | 18.81 |
| One coat primer, one coat undercoat and one coat gloss on pipes and gutters | | | | | | |
| exceeding 300mm girth | m2 | 0.78 | 12.48 | 1.96 | 2.17 | 16.61 |
| 150-300mm girth | m | 0.24 | 3.84 | 0.32 | 0.62 | 4.78 |

| | Unit | Labour hours | Labour cost £ | Materials £ | O & P £ | Total £ |
|---|---|---|---|---|---|---|
| **One coat primer, two coats undercoat and one coat gloss on metal general surfaces** | | | | | | |
| exceeding 300mm girth | m2 | 0.96 | 15.36 | 2.50 | 2.68 | 20.54 |
| not exceeding 150mm girth | m | 0.24 | 3.84 | 0.42 | 0.64 | 4.90 |
| 150-300mm girth | m | 0.32 | 5.12 | 0.84 | 0.89 | 6.85 |
| isolated areas not exceeding 0.5m2 | nr | 0.48 | 7.68 | 1.25 | 1.34 | 10.27 |
| **One coat primer, two coats undercoat and one coat gloss on metal windows, glazed doors and the like** | | | | | | |
| panes area not exceeding 0.1m2 | m2 | 1.60 | 25.60 | 0.98 | 3.99 | 30.57 |
| panes area 0.1 to 0.5m2 | m2 | 1.40 | 22.40 | 0.90 | 3.50 | 26.80 |
| panes area 0.5 to 1m2 | m2 | 1.20 | 19.20 | 0.82 | 3.00 | 23.02 |
| panes area exceeding 1m2 | m2 | 1.00 | 16.00 | 0.74 | 2.51 | 19.25 |
| **One coat primer, two coats undercoat and one coat gloss on structural steelwork** | | | | | | |
| exceeding 300mm girth | m2 | 1.20 | 19.20 | 2.50 | 3.26 | 24.96 |
| **One coat primer, two coats undercoat and one coat gloss on pipes and gutters** | | | | | | |
| exceeding 300mm girth | m2 | 1.04 | 16.64 | 2.50 | 2.87 | 22.01 |
| 150-300mm girth | m | 0.32 | 5.12 | 0.41 | 0.83 | 6.36 |

| | Unit | Labour hours | Labour cost £ | Materials £ | O & P £ | Total £ |
|---|---|---|---|---|---|---|
| **Wallpapering** | | | | | | |
| Prepare, size, apply adhesive, supply and hang paper to plastered walls and columns, butt jointed | | | | | | |
| lining paper | | | | | | |
| £1.50 per roll | m2 | 0.30 | 4.80 | 0.36 | 0.77 | 5.93 |
| £2.00 per roll | m2 | 0.30 | 4.80 | 0.48 | 0.79 | 6.07 |
| £2.50 per roll | m2 | 0.30 | 4.80 | 0.59 | 0.81 | 6.20 |
| washable paper | | | | | | |
| £2.50 per roll | m2 | 0.30 | 4.80 | 0.59 | 0.81 | 6.20 |
| £4.00 per roll | m2 | 0.30 | 4.80 | 0.94 | 0.86 | 6.60 |
| £5.00 per roll | m2 | 0.30 | 4.80 | 1.18 | 0.90 | 6.88 |
| vinyl paper | | | | | | |
| £4.00 per roll | m2 | 0.30 | 4.80 | 0.94 | 0.86 | 6.60 |
| £5.00 per roll | m2 | 0.30 | 4.80 | 1.18 | 0.90 | 6.88 |
| £6.00 per roll | m2 | 0.30 | 4.80 | 1.41 | 0.93 | 7.14 |
| washable paper | | | | | | |
| £5.00 per roll | m2 | 0.30 | 4.80 | 1.18 | 0.90 | 6.88 |
| £6.00 per roll | m2 | 0.30 | 4.80 | 1.41 | 0.93 | 7.14 |
| £7.00 per roll | m2 | 0.30 | 4.80 | 1.64 | 0.97 | 7.41 |
| hessian paper | | | | | | |
| £7.00 per m2 | m2 | 0.50 | 8.00 | 7.70 | 2.36 | 18.06 |
| £8.00 per m2 | m2 | 0.50 | 8.00 | 8.80 | 2.52 | 19.32 |
| £9.00 per m2 | m2 | 0.50 | 8.00 | 9.90 | 2.69 | 20.59 |
| suede paper | | | | | | |
| £9.00 per m2 | m2 | 0.30 | 4.80 | 9.90 | 2.21 | 16.91 |
| £10.00 per m2 | m2 | 0.30 | 4.80 | 11.00 | 2.37 | 18.17 |
| £11.00 per m2 | m2 | 0.30 | 4.80 | 12.10 | 2.54 | 19.44 |

| | Unit | Labour hours | Labour cost £ | Materials £ | O & P £ | Total £ |
|---|---|---|---|---|---|---|
| **Prepare, size, apply adhesive, supply and hang paper to plastered ceilings and columns, butt jointed** | | | | | | |
| | | | | | | |
| lining paper | | | | | | |
| £1.50 per roll | m2 | 0.35 | 5.60 | 0.36 | 0.89 | 6.85 |
| £2.00 per roll | m2 | 0.35 | 5.60 | 0.48 | 0.91 | 6.99 |
| £2.50 per roll | m2 | 0.35 | 5.60 | 0.59 | 0.93 | 7.12 |
| | | | | | | |
| washable paper | | | | | | |
| £2.50 per roll | m2 | 0.35 | 5.60 | 0.59 | 0.93 | 7.12 |
| £4.00 per roll | m2 | 0.35 | 5.60 | 0.94 | 0.98 | 7.52 |
| £5.00 per roll | m2 | 0.35 | 5.60 | 1.18 | 1.02 | 7.80 |
| | | | | | | |
| vinyl paper | | | | | | |
| £4.00 per roll | m2 | 0.35 | 5.60 | 0.94 | 0.98 | 7.52 |
| £5.00 per roll | m2 | 0.35 | 5.60 | 1.18 | 1.02 | 7.80 |
| £6.00 per roll | m2 | 0.35 | 5.60 | 1.41 | 1.05 | 8.06 |
| | | | | | | |
| washable paper | | | | | | |
| £5.00 per roll | m2 | 0.35 | 5.60 | 1.18 | 1.02 | 7.80 |
| £6.00 per roll | m2 | 0.35 | 5.60 | 1.41 | 1.05 | 8.06 |
| £7.00 per roll | m2 | 0.35 | 5.60 | 1.64 | 1.09 | 8.33 |

|  | Unit | Labour hours | Labour cost £ | Materials £ | O & P £ | Total £ |
|---|---|---|---|---|---|---|

## EXTERNAL WORKS

### Drainage

Excavate drain trench by hand, support sides, level and ram bottom of trench, backfill and consolidate with excavated material and remove surplus to skip for pipes 100mm and 150mm diameter, average depth of trench

| | Unit | Labour hours | Labour cost £ | Materials £ | O & P £ | Total £ |
|---|---|---|---|---|---|---|
| 0.50m | m | 1.20 | 16.80 | 0.00 | 2.52 | 19.32 |
| 0.75m | m | 1.90 | 26.60 | 0.00 | 3.99 | 30.59 |
| 1.00m | m | 2.50 | 35.00 | 0.00 | 5.25 | 40.25 |
| 1.25m | m | 4.00 | 56.00 | 0.00 | 8.40 | 64.40 |
| 1.50m | m | 5.00 | 70.00 | 0.00 | 10.50 | 80.50 |
| 1.75m | m | 6.00 | 84.00 | 0.00 | 12.60 | 96.60 |
| 2.00m | m | 7.20 | 100.80 | 0.00 | 15.12 | 115.92 |
| 2.25m | m | 8.20 | 114.80 | 0.00 | 17.22 | 132.02 |
| 2.50m | m | 9.50 | 133.00 | 0.00 | 19.95 | 152.95 |
| 2.75m | m | 10.50 | 147.00 | 0.00 | 22.05 | 169.05 |
| 3.00m | m | 12.00 | 168.00 | 0.00 | 25.20 | 193.20 |

Excavate drain trench by hand, support sides, level and ram bottom of trench, backfill and consolidate with excavated material and remove surplus to skip for pipes 225mm diameter, average depth of trench

| | Unit | Labour hours | Labour cost £ | Materials £ | O & P £ | Total £ |
|---|---|---|---|---|---|---|
| 0.50m | m | 1.30 | 18.20 | 0.00 | 2.73 | 20.93 |
| 0.75m | m | 2.10 | 29.40 | 0.00 | 4.41 | 33.81 |
| 1.00m | m | 2.70 | 37.80 | 0.00 | 5.67 | 43.47 |
| 1.25m | m | 4.30 | 60.20 | 0.00 | 9.03 | 69.23 |
| 1.50m | m | 5.40 | 75.60 | 0.00 | 11.34 | 86.94 |
| 1.75m | m | 6.60 | 92.40 | 0.00 | 13.86 | 106.26 |

| | Unit | Labour hours | Labour cost £ | Materials £ | O & P £ | Total £ |
|---|---|---|---|---|---|---|
| 2.00m | m | 7.10 | 99.40 | 0.00 | 14.91 | 114.31 |
| 2.25m | m | 9.00 | 126.00 | 0.00 | 18.90 | 144.90 |
| 2.50m | m | 10.20 | 142.80 | 0.00 | 21.42 | 164.22 |
| 2.75m | m | 11.30 | 158.20 | 0.00 | 23.73 | 181.93 |
| 3.00m | m | 12.80 | 179.20 | 0.00 | 26.88 | 206.08 |

Extra for breaking up by hand

| | Unit | Labour hours | Labour cost £ | Materials £ | O & P £ | Total £ |
|---|---|---|---|---|---|---|
| plain concrete 100mm thick | m2 | 0.90 | 12.60 | 0.00 | 1.89 | 14.49 |
| reinforced concrete 100mm thick | m2 | 1.00 | 14.00 | 0.00 | 2.10 | 16.10 |
| tarmacadam 75mm thick | m2 | 0.55 | 7.70 | 0.00 | 1.16 | 8.86 |
| hardcore 100mm thick | m2 | 0.40 | 5.60 | 0.00 | 0.84 | 6.44 |
| soft rock | m3 | 8.00 | 112.00 | 0.00 | 16.80 | 128.80 |
| hard rock | m3 | 10.00 | 140.00 | 0.00 | 21.00 | 161.00 |

Excavate drain trench by machine, support sides, level and ram bottom of trench, backfill and consolidate with excavated material and remove surplus to skip for pipes 100mm and 150mm diameter, average depth of trench

| | Unit | Labour hours | Labour cost £ | Plant £ | O & P £ | Total £ |
|---|---|---|---|---|---|---|
| 0.50m | m | 0.30 | 4.20 | 2.35 | 0.98 | 7.53 |
| 0.75m | m | 0.50 | 7.00 | 3.54 | 1.58 | 12.12 |
| 1.00m | m | 0.90 | 12.60 | 5.26 | 2.68 | 20.54 |
| 1.25m | m | 1.50 | 21.00 | 7.23 | 4.23 | 32.46 |
| 1.50m | m | 1.80 | 25.20 | 7.90 | 4.97 | 38.07 |
| 1.75m | m | 2.20 | 30.80 | 9.22 | 6.00 | 46.02 |
| 2.00m | m | 2.80 | 39.20 | 10.54 | 7.46 | 57.20 |
| 2.25m | m | 3.40 | 47.60 | 13.17 | 9.12 | 69.89 |
| 2.50m | m | 3.90 | 54.60 | 15.15 | 10.46 | 80.21 |
| 2.75m | m | 4.50 | 63.00 | 17.13 | 12.02 | 92.15 |
| 3.00m | m | 5.00 | 70.00 | 19.10 | 13.37 | 102.47 |

| | Unit | Labour hours | Labour cost £ | Plant £ | O & P £ | Total £ |
|---|---|---|---|---|---|---|
| **Excavate drain trench by machine, support sides, level and ram bottom of trench, backfill and consolidate with excavated material and remove surplus to skip for pipes 225mm diameter, average depth of trench** | | | | | | |
| 0.50m | m | 0.30 | 4.20 | 2.35 | 0.98 | 7.53 |
| 0.75m | m | 0.40 | 5.60 | 3.54 | 1.37 | 10.51 |
| 1.00m | m | 0.50 | 7.00 | 5.40 | 1.86 | 14.26 |
| 1.25m | m | 0.60 | 8.40 | 7.57 | 2.40 | 18.37 |
| 1.50m | m | 0.70 | 9.80 | 8.22 | 2.70 | 20.72 |
| 1.75m | m | 0.80 | 11.20 | 9.74 | 3.14 | 24.08 |
| 2.00m | m | 1.00 | 14.00 | 11.26 | 3.79 | 29.05 |
| 2.25m | m | 1.50 | 21.00 | 13.44 | 5.17 | 39.61 |
| 2.50m | m | 2.00 | 28.00 | 15.15 | 6.47 | 49.62 |
| 2.75m | m | 2.50 | 35.00 | 17.78 | 7.92 | 60.70 |
| 3.00m | m | 3.00 | 42.00 | 19.43 | 9.21 | 70.64 |
| **Extra for breaking up by machine** | | | | | | |
| plain concrete 100mm thick | m2 | 0.30 | 4.20 | 1.96 | 0.92 | 7.08 |
| reinforced concrete 100mm thick | m2 | 0.35 | 4.90 | 2.35 | 1.09 | 8.34 |
| tarmacadam 75mm thick | m2 | 0.20 | 2.80 | 1.44 | 0.64 | 4.88 |
| hardcore 100mm thick | m2 | 0.15 | 2.10 | 1.05 | 0.47 | 3.62 |
| soft rock | m3 | 4.55 | 63.70 | 7.63 | 10.70 | 82.03 |
| hard rock | m3 | 5.20 | 72.80 | 9.54 | 12.35 | 94.69 |

| | Unit | Labour hours | Labour cost £ | Materials £ | O & P £ | Total £ |
|---|---|---|---|---|---|---|
| **Sand bed in trench under pipe 100mm diameter, thickness** | | | | | | |
| 100mm | m | 0.10 | 1.40 | 1.24 | 0.40 | 3.04 |
| 150mm | m | 0.12 | 1.68 | 1.83 | 0.53 | 4.04 |

|  | Unit | Labour hours | Labour cost £ | Materials £ | O & P £ | Total £ |
|---|---|---|---|---|---|---|
| **Sand bed in trench under pipe 150mm diameter, thickness** | | | | | | |
| 100mm | m | 0.11 | 1.54 | 1.70 | 0.49 | 3.73 |
| 150mm | m | 0.13 | 1.82 | 1.89 | 0.56 | 4.27 |
| **Sand bed in trench under pipe 225mm diameter, thickness** | | | | | | |
| 100mm | m | 0.14 | 1.96 | 2.03 | 0.60 | 4.59 |
| 150mm | m | 0.16 | 2.24 | 2.94 | 0.78 | 5.96 |
| **Granular filling in bed in trench under pipe 100mm diameter, thickness** | | | | | | |
| 100mm | m | 0.12 | 1.68 | 1.05 | 0.41 | 3.14 |
| 150mm | m | 0.14 | 1.96 | 1.51 | 0.52 | 3.99 |
| **Granular filling in bed in trench under pipe 150mm diameter, thickness** | | | | | | |
| 100mm | m | 0.13 | 1.82 | 1.16 | 0.45 | 3.43 |
| 150mm | m | 0.15 | 2.10 | 1.76 | 0.58 | 4.44 |
| **Granular filling in bed in trench under pipe 225mm diameter, thickness** | | | | | | |
| 100mm | m | 0.16 | 2.24 | 1.50 | 0.56 | 4.30 |
| 150mm | m | 0.18 | 2.52 | 2.30 | 0.72 | 5.54 |
| **Concrete in bed in trench under pipe, 100mm diameter, thickness** | | | | | | |
| 100mm | m | 0.24 | 3.36 | 4.60 | 1.19 | 9.15 |
| 150mm | m | 0.28 | 3.92 | 6.85 | 1.62 | 12.39 |

| | Unit | Labour hours | Labour cost £ | Materials £ | O & P £ | Total £ |
|---|---|---|---|---|---|---|
| **Concrete in bed in trench under pipe, 150mm diameter, thickness** | | | | | | |
| 100mm | m | 0.26 | 3.64 | 5.32 | 1.34 | 10.30 |
| 150mm | m | 0.30 | 4.20 | 6.77 | 1.65 | 12.62 |
| **Concrete in bed in trench under pipe, 225mm diameter, thickness** | | | | | | |
| 100mm | m | 0.32 | 4.48 | 6.05 | 1.58 | 12.11 |
| 150mm | m | 0.36 | 5.04 | 9.28 | 2.15 | 16.47 |
| **Concrete in bed and haunching to pipe, 100mm diameter, bed thickness** | | | | | | |
| 100mm | m | 0.48 | 6.72 | 8.42 | 2.27 | 17.41 |
| 150mm | m | 0.56 | 7.84 | 12.59 | 3.06 | 23.49 |
| **Concrete in bed and haunching to pipe, 150mm diameter, bed thickness** | | | | | | |
| 100mm | m | 0.52 | 7.28 | 9.52 | 2.52 | 19.32 |
| 150mm | m | 0.60 | 8.40 | 13.01 | 3.21 | 24.62 |
| **Concrete in bed and haunching to pipe, 225mm diameter, bed thickness** | | | | | | |
| 100mm | m | 0.60 | 8.40 | 10.72 | 2.87 | 21.99 |
| 150mm | m | 0.66 | 9.24 | 13.88 | 3.47 | 26.59 |
| **Granular filling in bed and surround to pipe, 100mm diameter, thickness** | | | | | | |
| 100mm | m | 0.36 | 5.04 | 4.22 | 1.39 | 10.65 |
| 150mm | m | 0.42 | 5.88 | 6.53 | 1.86 | 14.27 |

| | Unit | Labour hours | Labour cost £ | Materials £ | O & P £ | Total £ |
|---|---|---|---|---|---|---|
| **Granular filling in bed and surround to pipe, 150mm diameter, thickness** | | | | | | |
| 100mm | m | 0.40 | 5.60 | 5.42 | 1.65 | 12.67 |
| 150mm | m | 0.46 | 6.44 | 8.20 | 2.20 | 16.84 |
| **Granular filling in bed and surround to pipe, 225mm diameter, thickness** | | | | | | |
| 100mm | m | 0.44 | 6.16 | 8.14 | 2.15 | 16.45 |
| 150mm | m | 0.50 | 7.00 | 9.73 | 2.51 | 19.24 |
| **Concrete in bed and surround to pipe, 100mm diameter, thickness** | | | | | | |
| 100mm | m | 0.72 | 10.08 | 9.10 | 2.88 | 22.06 |
| 150mm | m | 0.84 | 11.76 | 9.90 | 3.25 | 24.91 |
| **Concrete in bed and surround to pipe, 150mm diameter, thickness** | | | | | | |
| 100mm | m | 0.72 | 10.08 | 9.89 | 3.00 | 22.97 |
| 150mm | m | 0.84 | 11.76 | 10.81 | 3.39 | 25.96 |
| **Concrete in bed and surround to pipe, 225mm diameter, thickness** | | | | | | |
| 100mm | m | 0.72 | 10.08 | 12.79 | 3.43 | 26.30 |
| 150mm | m | 0.84 | 11.76 | 15.07 | 4.02 | 30.85 |

| | Unit | Labour hours | Labour cost £ | Materials £ | O & P £ | Total £ |
|---|---|---|---|---|---|---|
| Hepworths' Supersleve vitrified clay drain pipes, spigot and socket joints with sealing rings, 100mm diameter | | | | | | |
| laid in trenches | m | 0.36 | 5.76 | 7.60 | 2.00 | 15.36 |
| in lengths not exceeding 3m | m | 0.54 | 8.64 | 7.60 | 2.44 | 18.68 |
| bend | nr | 0.30 | 4.80 | 10.32 | 2.27 | 17.39 |
| rest bend | nr | 0.30 | 4.80 | 16.07 | 3.13 | 24.00 |
| single junction | nr | 0.30 | 4.80 | 15.15 | 2.99 | 22.94 |
| Hepworths' Supersleve vitrified clay drain pipes, spigot and socket joints with sealing rings, 150mm diameter | | | | | | |
| laid in trenches | m | 0.60 | 9.60 | 14.47 | 3.61 | 27.68 |
| in lengths not exceeding 3m | m | 0.40 | 6.40 | 17.77 | 3.63 | 27.80 |
| bend | nr | 0.30 | 4.80 | 14.44 | 2.89 | 22.13 |
| rest bend | nr | 0.30 | 4.80 | 18.56 | 3.50 | 26.86 |
| single junction | nr | 0.30 | 4.80 | 19.35 | 3.62 | 27.77 |
| Vitrified clay inlet gulley complete with grid and surrounded in concrete | nr | 1.50 | 24.00 | 48.20 | 10.83 | 83.03 |
| Vitrified clay back inlet gulley complete with grid and surrounded in concrete | nr | 1.50 | 24.00 | 68.88 | 13.93 | 106.81 |
| Vitrified clay paved area gulley complete with grid and surrounded in concrete | nr | 1.50 | 24.00 | 56.03 | 12.00 | 92.03 |

| | Unit | Labour hours | Labour cost £ | Materials £ | O & P £ | Total £ |
|---|---|---|---|---|---|---|
| **Manholes** | | | | | | |
| Excavate by hand for manhole not exceeding | | | | | | |
| 1.0m deep | m3 | 4.00 | 64.00 | 0.00 | 9.60 | 73.60 |
| 1.5m deep | m3 | 4.50 | 72.00 | 0.00 | 10.80 | 82.80 |
| 2.0m deep | m3 | 5.60 | 89.60 | 0.00 | 13.44 | 103.04 |
| Dispose surplus excavated material by hand, load into barrows, wheel 50m and deposit into skip | m3 | 2.20 | 35.20 | 0.00 | 5.28 | 40.48 |
| Excavate by machine for manhole not exceeding | | | | | | |
| 1.0m deep | m3 | 0.35 | 5.60 | 5.28 | 1.63 | 12.51 |
| 1.5m deep | m3 | 0.40 | 6.40 | 5.53 | 1.79 | 13.72 |
| 2.0m deep | m3 | 0.45 | 7.20 | 5.77 | 1.95 | 14.92 |
| Earthwork support not exceeding 2m between opposing faces, depth not exceeding | | | | | | |
| 1.0m deep | m2 | 0.35 | 5.60 | 1.62 | 1.08 | 8.30 |
| 2.0m deep | m2 | 0.40 | 6.40 | 1.75 | 1.22 | 9.37 |
| 4.0m deep | m2 | 0.45 | 7.20 | 1.06 | 1.24 | 9.50 |
| Site-mixed concrete in base of manhole, thickness | | | | | | |
| 100-150mm | m3 | 2.00 | 32.00 | 78.21 | 16.53 | 126.74 |
| 150-300mm | m3 | 1.80 | 28.80 | 78.21 | 16.05 | 123.06 |
| Site-mixed concrete in benching to manhole, average thickness | | | | | | |
| 225mm | m3 | 6.00 | 96.00 | 87.43 | 27.51 | 210.94 |

|  | Unit | Labour hours | Labour cost £ | Materials £ | O & P £ | Total £ |
|---|---|---|---|---|---|---|
| Engineering bricks Class 'B' in cement mortar in walls of manhole | m2 | 4.80 | 76.80 | 59.81 | 20.49 | 157.10 |
| Extra for fair face and flush pointing | m2 | 0.25 | 4.00 | 0.00 | 0.60 | 4.60 |
| Build in ends of pipes to one brick wall and make good, pipe diameter |  |  |  |  |  |  |
| 100mm | nr | 0.15 | 2.40 | 0.00 | 0.36 | 2.76 |
| 150mm | nr | 0.19 | 3.02 | 0.00 | 0.45 | 3.48 |
| Galvanised step iron built into brickwork | nr | 0.20 | 3.20 | 5.96 | 1.37 | 10.53 |
| Cast iron manhole cover, frame bedded in cement mortar |  |  |  |  |  |  |
| Grade A, light duty, size 600 × 450mm | nr | 1.90 | 30.40 | 38.42 | 10.32 | 79.14 |
| Grade B, medium duty, size 600 × 450mm | nr | 2.00 | 32.00 | 122.28 | 23.14 | 177.42 |
| Best quality vitrified clay channels, bedded in cement mortar |  |  |  |  |  |  |
| half-section straight main channel |  |  |  |  |  |  |
| 100mm dia. × 300mm | nr | 0.75 | 12.00 | 3.17 | 2.28 | 17.45 |
| 100mm dia. × 600mm | nr | 0.85 | 13.60 | 3.17 | 2.52 | 19.29 |
| 100mm dia. × 1000mm | nr | 0.95 | 15.20 | 4.43 | 2.94 | 22.57 |
| 150mm dia. × 300mm | nr | 0.75 | 12.00 | 5.34 | 2.60 | 19.94 |
| 150mm dia. × 1000mm | nr | 0.85 | 13.60 | 5.34 | 2.84 | 21.78 |
| 150mm dia. × 1000mm | nr | 0.95 | 15.20 | 7.81 | 3.45 | 26.46 |
| Half-section 90 degrees channel bend |  |  |  |  |  |  |
| 100mm | nr | 0.30 | 4.80 | 4.50 | 1.40 | 10.70 |
| 150mm | nr | 0.40 | 6.40 | 7.77 | 2.13 | 16.30 |

|  | Unit | Labour hours | Labour cost £ | Materials £ | O & P £ | Total £ |
|---|---|---|---|---|---|---|
| **Three-quarter section 90 degrees channel bend** | | | | | | |
| 100mm | nr | 1.20 | 19.20 | 12.70 | 4.79 | 36.69 |
| 150mm | nr | 1.30 | 20.80 | 22.09 | 6.43 | 49.32 |

|  | Unit | Specialist price £ | O & P £ | Total £ |
|---|---|---|---|---|
| **Fencing** | | | | |
| Chainlink fencing, galvanised steel mesh on three strained line wires fixed to concrete posts at 3000mm centres | | | | |
| 900mm high | m | 17.09 | 2.56 | 19.65 |
| 1200mm high | m | 20.64 | 3.10 | 23.74 |
| 1400mm high | m | 23.21 | 3.48 | 26.69 |
| 1800mm high | m | 30.94 | 4.64 | 35.58 |
| Chainlink fencing, galvanised steel mesh on three strained line wires fixed to galvanised mild steel posts at 3000mm centres | | | | |
| 900mm high | m | 18.12 | 2.72 | 20.84 |
| 1200mm high | m | 21.43 | 3.21 | 24.64 |
| 1400mm high | m | 25.31 | 3.80 | 29.11 |
| 1800mm high | m | 32.92 | 4.94 | 37.86 |
| Chainlink fencing, galvanised mild steel mesh on three strained line wires fixed to concrete posts at 3000mm centres | | | | |
| 900mm high | m | 17.26 | 2.59 | 19.85 |
| 1200mm high | m | 21.23 | 3.18 | 24.41 |
| 1400mm high | m | 23.75 | 3.56 | 27.31 |
| 1800mm high | m | 31.35 | 4.70 | 36.05 |

| | Unit | Specialist price £ | O & P £ | Total £ |
|---|---|---|---|---|
| Chainlink fencing, plastic-coated mild steel mesh on three strained line wires fixed to galvanised mild steel posts at 3000mm centres | | | | |
| 900mm high | m | 18.76 | 2.81 | 21.57 |
| 1200mm high | m | 22.03 | 3.30 | 25.33 |
| 1400mm high | m | 27.00 | 4.05 | 31.05 |
| 1800mm high | m | 33.30 | 5.00 | 38.30 |
| Strained wire fence including concrete posts at 3000mm centres | | | | |
| 1.00m high, 5 wires | m | 14.07 | 2.11 | 16.18 |
| 1.20m high, 6 wires | m | 14.46 | 2.17 | 16.63 |
| 1.40m high, 8 wires | m | 15.24 | 2.29 | 17.53 |
| | | | 0.00 | 0.00 |
| Cleft chestnut fencing, pales set 75mm apart, 75mm diameter softwood posts at 2500mm centres | | | | |
| 0.90m high, 2 wires | m | 12.67 | 1.90 | 14.57 |
| 1.05m high, 2 wires | m | 13.00 | 1.95 | 14.95 |
| 1.20m high, 2 wires | m | 13.66 | 2.05 | 15.71 |
| 1.50m high, 3 wires | m | 14.92 | 2.24 | 17.16 |
| 1.80m high, 4 wires | m | 16.13 | 2.42 | 18.55 |
| Close-boarded fencing consisting of 90 x 19mm pales lapped 13mm fixed to 75 x 38mm horizontal rails on 75 x 75mm posts at 3000mm centres, height | | | | |
| 1.00m | m | 40.44 | 6.07 | 46.51 |
| 1.20m | m | 41.32 | 6.20 | 47.52 |
| 1.40m | m | 42.39 | 6.36 | 48.75 |
| 1.60m | m | 46.27 | 6.94 | 53.21 |
| 1.80m | m | 48.21 | 7.23 | 55.44 |

| | Unit | Labour hours | Labour cost £ | Materials £ | O & P £ | Total £ |
|---|---|---|---|---|---|---|

### Kerbs and edgings

Excavate trench by hand for kerb foundation

| | Unit | Labour hours | Labour cost £ | Materials £ | O & P £ | Total £ |
|---|---|---|---|---|---|---|
| 200 × 75mm | m | 0.10 | 1.40 | 0.00 | 0.21 | 1.61 |
| 250 × 100mm | m | 0.12 | 1.68 | 0.00 | 0.25 | 1.93 |
| 300 × 100mm | m | 0.14 | 1.96 | 0.00 | 0.29 | 2.25 |
| 450 × 150mm | m | 0.20 | 2.80 | 0.00 | 0.42 | 3.22 |
| 600 × 200mm | m | 0.40 | 5.60 | 0.00 | 0.84 | 6.44 |

Excavate curved trench by hand for kerb foundation

| | Unit | Labour hours | Labour cost £ | Materials £ | O & P £ | Total £ |
|---|---|---|---|---|---|---|
| 250 × 100mm | m | 0.12 | 1.68 | 0.00 | 0.25 | 1.93 |
| 250 × 100mm | m | 0.14 | 1.96 | 0.00 | 0.29 | 2.25 |
| 300 × 100mm | m | 0.16 | 2.24 | 0.00 | 0.34 | 2.58 |
| 600 × 200mm | m | 0.24 | 3.36 | 0.00 | 0.50 | 3.86 |
| 600 × 200mm | m | 0.44 | 6.16 | 0.00 | 0.92 | 7.08 |

| | Unit | Labour hours | Labour cost £ | Plant £ | O & P £ | Total £ |
|---|---|---|---|---|---|---|

Excavate trench by machine for kerb foundation
200 × 75mm

| | Unit | Labour hours | Labour cost £ | Plant £ | O & P £ | Total £ |
|---|---|---|---|---|---|---|
| 200 × 75mm | m | 0.00 | 0.00 | 0.77 | 0.12 | 0.89 |
| 250 × 100mm | m | 0.00 | 0.00 | 0.83 | 0.12 | 0.95 |
| 600  200mm | m | 0.00 | 0.00 | 0.89 | 0.13 | 1.02 |
| 450 × 150mm | m | 0.00 | 0.00 | 1.36 | 0.20 | 1.56 |
| 600 × 200mm | m | 0.00 | 0.00 | 2.63 | 0.39 | 3.02 |

Excavate curved trench by machine for kerb foundation

| | Unit | Labour hours | Labour cost £ | Plant £ | O & P £ | Total £ |
|---|---|---|---|---|---|---|
| 200 × 75mm | m | 0.00 | 0.00 | 0.89 | 0.13 | 1.02 |
| 250 × 100mm | m | 0.00 | 0.00 | 0.95 | 0.14 | 1.09 |
| 300 × 100mm | m | 0.00 | 0.00 | 1.01 | 0.15 | 1.16 |
| 450 × 150mm | m | 0.00 | 0.00 | 1.74 | 0.26 | 2.00 |
| 600 × 200mm | m | 0.00 | 0.00 | 3.07 | 0.46 | 3.53 |

|  | Unit | Labour hours | Labour cost £ | Materials £ | O & P £ | Total £ |
|---|---|---|---|---|---|---|
| **Site-mixed concrete in foundation for kerb size** | | | | | | |
| 200 × 75mm | m | 0.03 | 0.42 | 1.23 | 0.25 | 1.90 |
| 250 × 100mm | m | 0.04 | 0.56 | 2.03 | 0.39 | 2.98 |
| 300 × 100mm | m | 0.05 | 0.70 | 2.44 | 0.47 | 3.61 |
| 450 × 150mm | m | 0.18 | 2.52 | 5.53 | 1.21 | 9.26 |
| 600 × 200mm | m | 0.30 | 4.20 | 9.81 | 2.10 | 16.11 |
| **Precast concrete kerbs, channels and edgings, jointed and pointed in cement mortar** | | | | | | |
| kerbs, straight | | | | | | |
| 127 × 254mm | m | 0.40 | 5.60 | 10.86 | 2.47 | 18.93 |
| 152 × 305mm | m | 0.45 | 6.30 | 12.40 | 2.81 | 21.51 |
| kerbs, curved | | | | | | |
| 127 × 254mm | m | 0.50 | 7.00 | 15.05 | 3.31 | 25.36 |
| 152 × 305mm | m | 0.55 | 7.70 | 17.06 | 3.71 | 28.47 |
| channels, straight | | | | | | |
| 127 × 254mm | m | 0.40 | 5.60 | 11.82 | 2.61 | 20.03 |
| channels, curved | | | | | | |
| 127 × 254mm | m | 0.50 | 7.00 | 16.10 | 3.47 | 26.57 |
| edgings, straight | | | | | | |
| 51 × 152mm | m | 0.30 | 4.20 | 8.02 | 1.83 | 14.05 |
| 51 × 203mm | m | 0.30 | 4.20 | 9.07 | 1.99 | 15.26 |

**Sub-bases**

Beds and bases compacting in layers and grading

| average thickness, 100mm | | | | | | |
|---|---|---|---|---|---|---|
| granular fill | m2 | 0.10 | 1.40 | 3.42 | 0.72 | 5.54 |
| sand | m2 | 0.12 | 1.68 | 3.15 | 0.72 | 5.55 |
| hardcore | m2 | 0.13 | 1.82 | 3.02 | 0.73 | 5.57 |

| | Unit | Labour hours | Labour cost £ | Materials £ | O & P £ | Total £ |
|---|---|---|---|---|---|---|
| **average thickness, 150mm** | | | | | | |
| granular fill | m2 | 0.12 | 1.68 | 4.25 | 0.89 | 6.82 |
| sand | m2 | 0.14 | 1.96 | 4.08 | 0.91 | 6.95 |
| hardcore | m2 | 0.16 | 2.24 | 3.88 | 0.92 | 7.04 |
| **average thickness, 200mm** | | | | | | |
| granular fill | m2 | 0.14 | 1.96 | 5.53 | 1.12 | 8.61 |
| sand | m2 | 0.16 | 2.24 | 4.99 | 1.08 | 8.31 |
| hardcore | m2 | 0.18 | 2.52 | 4.80 | 1.10 | 8.42 |

## Concrete beds

Site-mixed concrete in beds

| | Unit | Labour hours | Labour cost £ | Materials £ | O & P £ | Total £ |
|---|---|---|---|---|---|---|
| **not exceeding 150mm** | | | | | | |
| thick | m3 | 2.80 | 39.20 | 91.59 | 19.62 | 150.41 |
| 150 to 450mm thick | m3 | 2.20 | 30.80 | 91.59 | 18.36 | 140.75 |

Formwork to sides of concrete bases

| | Unit | Labour hours | Labour cost £ | Materials £ | O & P £ | Total £ |
|---|---|---|---|---|---|---|
| **not exceeding 250mm** | | | | | | |
| wide | m | 0.65 | 9.10 | 2.12 | 1.68 | 12.90 |
| 250 to 500mm wide | m | 0.90 | 12.60 | 3.47 | 2.41 | 18.48 |

Steel fabric reinforcement laid in concrete beds

| | Unit | Labour hours | Labour cost £ | Materials £ | O & P £ | Total £ |
|---|---|---|---|---|---|---|
| ref A142, 2.22kg/m2 | m2 | 0.15 | 2.10 | 1.61 | 0.56 | 4.27 |
| ref A193, 3.02kg/m2 | m2 | 0.18 | 2.52 | 1.95 | 0.67 | 5.14 |

Expansion joint, impregnated fibre-base joint filler, formed joint

| | Unit | Labour hours | Labour cost £ | Materials £ | O & P £ | Total £ |
|---|---|---|---|---|---|---|
| **12.5mm thick** | | | | | | |
| **not exceeding 150mm** | | | | | | |
| wide | m | 0.18 | 2.52 | 2.76 | 0.79 | 6.07 |
| 150 to 300mm wide | m | 0.24 | 3.36 | 4.24 | 1.14 | 8.74 |
| 300 to 450mm wide | m | 0.30 | 4.20 | 6.41 | 1.59 | 12.20 |

|  | Unit | Labour hours | Labour cost £ | Materials £ | O & P £ | Total £ |
|---|---|---|---|---|---|---|
| **25mm thick** | | | | | | |
| not exceeding 150mm wide | m | 0.20 | 2.80 | 3.77 | 0.99 | 7.56 |
| 150 to 300mm wide | m | 0.26 | 3.64 | 5.61 | 1.39 | 10.64 |
| 300 to 450mm wide | m | 0.32 | 4.48 | 8.72 | 1.51 | 14.71 |
| **Treat surfaces of concrete before setting** | | | | | | |
| tamping | m2 | 0.06 | 0.84 | 0.00 | 0.13 | 0.97 |
| floating | m2 | 0.10 | 1.40 | 0.00 | 0.21 | 1.61 |
| trowelling | m2 | 0.15 | 2.10 | 0.00 | 0.32 | 2.42 |

|  | Unit | Specialist price £ | O & P £ | Total £ |
|---|---|---|---|---|
| **Pavings** | | | | |
| Bitumen macadam paving consisting of 50mm base course with 28mm aggregate and 20mm thick wearing course with 6mm aggregate | m2 | 22.73 | 3.41 | 26.14 |
| Bitumen macadam paving consisting of 70mm base course with 40mm aggregate and 25mm thick wearing course with 10mm aggregate | m2 | 27.16 | 4.07 | 31.23 |

|  | Unit | Labour hours | Labour cost £ | Materials £ | O & P £ | Total £ |
|---|---|---|---|---|---|---|
| **Gravel paving in two layers, bottom layer clinker aggregate, second layer fine gravel aggregate** | | | | | | |
| 50mm thick | m | 0.10 | 1.40 | 2.90 | 0.21 | 4.51 |
| 75mm thick | m | 0.12 | 1.68 | 3.78 | 0.69 | 6.15 |
| 100mm thick | m | 0.14 | 1.96 | 5.26 | 0.86 | 8.08 |

| | Unit | Labour hours | Labour cost £ | Materials £ | O & P £ | Total £ |
|---|---|---|---|---|---|---|
| **Brick paving, 215 x 103 x 65mm laid to falls and cross falls bedding in cement mortar 15mm thick** | | | | | | |
| straight joints both ways | | | | | | |
| bricks laid flat | m2 | 1.50 | 24.00 | 25.69 | 3.60 | 53.29 |
| bricks laid on edge | m2 | 1.75 | 28.00 | 39.15 | 8.05 | 75.20 |
| herringbone pattern | | | | | | |
| bricks laid flat | m2 | 1.75 | 28.00 | 25.69 | 4.20 | 57.89 |
| bricks laid on edge | m2 | 2.00 | 32.00 | 39.15 | 8.65 | 79.80 |
| **Preacast concrete paving flags spot bedded in cement mortar** | | | | | | |
| natural colour | | | | | | |
| 450 × 450 × 50mm | m2 | 0.50 | 8.00 | 14.78 | 1.20 | 23.98 |
| 600 × 600 × 50mm | m2 | 0.48 | 7.68 | 12.13 | 3.37 | 23.18 |
| 600 × 900 × 50mm | m2 | 0.45 | 7.20 | 11.26 | 2.90 | 21.36 |
| coloured | | | | | | |
| 450 × 450 × 50mm | m2 | 0.50 | 8.00 | 16.37 | 1.20 | 25.57 |
| 600 × 600 × 50mm | m2 | 0.48 | 7.68 | 13.14 | 3.61 | 24.43 |
| 600 × 900 × 50mm | m2 | 0.15 | 2.40 | 12.05 | 2.33 | 16.78 |

| | Unit | Specialist price £ | O & P £ | Total £ |
|---|---|---|---|---|

## ALTERATIONS AND REPAIRS

### Shoring

| | Unit | Specialist price £ | O & P £ | Total £ |
|---|---|---|---|---|
| Raking shoring consisting of two 175 × 175mm rakers, each 5m length, 50 × 150mm wall plate including sole pieces, steel dogs needles, cleats and bracing | nr | 835.05 | 125.26 | 960.31 |
| Raking shoring consisting of two 225 × 225mm rakers, each 8m length, 50 × 200mm wall plate including sole pieces, steel dogs needles, cleats, bracing and cutting through walls, floors and ceilings | nr | 1049.48 | 157.42 | 1206.90 |
| Dead shoring consisting of two 175 × 175mm legs including needle, sole plate, braces, wedges steel dogs and cutting through walls, floors and ceilings | nr | 730.95 | 109.64 | 840.59 |
| Flying shoring consisting of 100 × 150mm shores, 50 × 200 wall plates, 50 × 100mm struts and straining pieces, distance between walls being supported | | | | |
| 3m | nr | 620.54 | 93.08 | 713.62 |
| 4m | nr | 663.70 | 99.56 | 763.26 |
| 5m | nr | 730.95 | 109.64 | 840.59 |
| Strut existing window openings consisting of 125 × 50mm wall plates and struts | nr | 147.89 | 22.18 | 170.07 |

| | Unit | Labour hours | Labour cost £ | Materials £ | O & P £ | Total £ |
|---|---|---|---|---|---|---|

## Forming openings

Form opening for windows or
doors in existing walls in cement
mortar and make good to sides
of openings

| | Unit | Labour hours | Labour cost £ | Materials £ | O & P £ | Total £ |
|---|---|---|---|---|---|---|
| 75mm blockwork | m2 | 2.00 | 28.00 | 2.88 | 4.63 | 35.51 |
| 100mm blockwork | m2 | 2.22 | 31.08 | 2.88 | 5.09 | 39.05 |
| 140mm blockwork | m2 | 2.65 | 37.10 | 3.63 | 6.11 | 46.84 |
| 215mm blockwork | m2 | 2.80 | 39.20 | 4.24 | 6.52 | 49.96 |
| half brick wall | m2 | 2.90 | 40.60 | 2.88 | 6.52 | 50.00 |
| one brick wall | m2 | 3.78 | 52.92 | 3.69 | 8.49 | 65.10 |
| one and a half brick wall | m2 | 4.80 | 67.20 | 4.59 | 10.77 | 82.56 |
| two brick wall | m2 | 7.20 | 100.80 | 6.03 | 16.02 | 122.85 |

Form opening for windows or
doors in existing walls in lime
mortar and make good to sides
of openings

| | Unit | Labour hours | Labour cost £ | Materials £ | O & P £ | Total £ |
|---|---|---|---|---|---|---|
| 75mm blockwork | m2 | 1.80 | 25.20 | 2.88 | 4.21 | 32.29 |
| 100mm blockwork | m2 | 2.00 | 28.00 | 2.88 | 4.63 | 35.51 |
| 140mm blockwork | m2 | 2.40 | 33.60 | 3.63 | 5.58 | 42.81 |
| 215mm blockwork | m2 | 2.55 | 35.70 | 4.24 | 5.99 | 45.93 |
| half brick wall | m2 | 2.60 | 36.40 | 2.88 | 5.89 | 45.17 |
| one brick wall | m2 | 3.40 | 47.60 | 3.69 | 7.69 | 58.98 |
| one and a half brick wall | m2 | 4.80 | 67.20 | 4.59 | 10.77 | 82.56 |
| two brick wall | m2 | 6.50 | 91.00 | 6.03 | 14.55 | 111.58 |

Form opening for lintels above
openings in existing walls in cement
mortar and make good to sides
of openings

| | Unit | Labour hours | Labour cost £ | Materials £ | O & P £ | Total £ |
|---|---|---|---|---|---|---|
| 75mm blockwork | m2 | 2.98 | 41.72 | 1.27 | 6.45 | 49.44 |
| 100mm blockwork | m2 | 3.36 | 47.04 | 1.27 | 7.25 | 55.56 |
| 140mm blockwork | m2 | 3.87 | 54.18 | 2.05 | 8.43 | 64.66 |
| 215mm blockwork | m2 | 4.23 | 59.22 | 2.82 | 9.31 | 71.35 |
| half brick wall | m2 | 6.30 | 88.20 | 2.05 | 13.54 | 103.79 |

| | Unit | Labour hours | Labour cost £ | Materials £ | O & P £ | Total £ |
|---|---|---|---|---|---|---|
| one brick wall | m2 | 10.50 | 147.00 | 3.69 | 22.60 | 173.29 |
| one and a half brick wall | m2 | 12.60 | 176.40 | 4.59 | 27.15 | 208.14 |
| two brick wall | m2 | 16.98 | 237.72 | 6.03 | 36.56 | 280.31 |

Form opening for lintels above openings in existing walls in lime mortar and make good to sides of openings

| | Unit | Labour hours | Labour cost £ | Materials £ | O & P £ | Total £ |
|---|---|---|---|---|---|---|
| 75mm blockwork | m2 | 2.75 | 38.50 | 1.27 | 5.97 | 45.74 |
| 100mm blockwork | m2 | 3.00 | 42.00 | 1.27 | 6.49 | 49.76 |
| 140mm blockwork | m2 | 3.45 | 48.30 | 2.05 | 7.55 | 57.90 |
| 215mm blockwork | m2 | 3.86 | 54.04 | 2.82 | 8.53 | 65.39 |
| half brick wall | m2 | 5.75 | 80.50 | 2.05 | 12.38 | 94.93 |
| one brick wall | m2 | 9.45 | 132.30 | 3.69 | 20.40 | 156.39 |
| one and a half brick wall | m2 | 11.40 | 159.60 | 4.59 | 24.63 | 188.82 |
| two brick wall | m2 | 15.30 | 214.20 | 6.03 | 33.03 | 253.26 |

Form opening in reinforced concrete floor slab and make good to edges, slab thickness

| | Unit | Labour hours | Labour cost £ | Materials £ | O & P £ | Total £ |
|---|---|---|---|---|---|---|
| 100mm | m2 | 5.65 | 79.10 | 3.21 | 12.35 | 94.66 |
| 150mm | m2 | 7.86 | 110.04 | 3.53 | 17.04 | 130.61 |
| 200mm | m2 | 9.50 | 133.00 | 3.85 | 20.53 | 157.38 |
| 250mm | m2 | 11.57 | 161.98 | 4.17 | 24.92 | 191.07 |
| 300mm | m2 | 14.55 | 203.70 | 7.72 | 31.71 | 243.13 |

Form opening in reinforced concrete walls and make good to edges, wall thickness

| | Unit | Labour hours | Labour cost £ | Materials £ | O & P £ | Total £ |
|---|---|---|---|---|---|---|
| 100mm | m2 | 6.50 | 91.00 | 3.21 | 14.13 | 108.34 |
| 150mm | m2 | 7.20 | 100.80 | 3.53 | 15.65 | 119.98 |
| 200mm | m2 | 10.33 | 144.62 | 3.85 | 22.27 | 170.74 |
| 250mm | m2 | 12.10 | 169.40 | 4.17 | 26.04 | 199.61 |
| 300mm | m2 | 15.32 | 214.48 | 7.72 | 33.33 | 255.53 |

| | Unit | Labour hours | Labour cost £ | Materials £ | O & P £ | Total £ |
|---|---|---|---|---|---|---|

**Filling openings**

Take out existing door and frame
complete, make good to reveals
and piece up skirting to match
existing

| | Unit | Labour hours | Labour cost £ | Materials £ | O & P £ | Total £ |
|---|---|---|---|---|---|---|
| single internal door | nr | 6.00 | 84.00 | 12.71 | 14.51 | 111.22 |
| single external door | nr | 6.20 | 86.80 | 12.71 | 14.93 | 114.44 |
| double internal door | nr | 6.20 | 86.80 | 13.43 | 15.03 | 115.26 |
| double external door | nr | 6.40 | 89.60 | 13.43 | 15.45 | 118.48 |

Take out existing single door and
frame complete, fill in opening, fix
new skirting to match existing and
plaster both sides

| | Unit | Labour hours | Labour cost £ | Materials £ | O & P £ | Total £ |
|---|---|---|---|---|---|---|
| 100mm blockwork | m2 | 10.45 | 146.30 | 46.11 | 28.86 | 221.27 |
| 215mm blockwork | m2 | 12.47 | 174.58 | 67.41 | 36.30 | 278.29 |
| half brick wall | m2 | 12.40 | 173.60 | 42.12 | 32.36 | 248.08 |
| one brick wall | m2 | 14.60 | 204.40 | 75.98 | 42.06 | 322.44 |
| one and a half brick wall | m2 | 16.37 | 229.18 | 104.69 | 50.08 | 383.95 |

Take out existing double doors and
frame complete, fill in opening, fix
new skirting to match existing and
plaster both sides

| | Unit | Labour hours | Labour cost £ | Materials £ | O & P £ | Total £ |
|---|---|---|---|---|---|---|
| 100mm blockwork | m2 | 14.59 | 204.26 | 54.37 | 38.79 | 297.42 |
| 215mm blockwork | m2 | 16.36 | 229.04 | 122 | 52.66 | 403.70 |
| half brick wall | m2 | 15.54 | 217.56 | 70.05 | 43.14 | 330.75 |
| one brick wall | m2 | 20.34 | 284.76 | 128.68 | 62.02 | 475.46 |
| one and a half brick wall | m2 | 22.20 | 310.80 | 105.09 | 62.38 | 478.27 |

| | Unit | Labour hours | Labour cost £ | Materials £ | O & P £ | Total £ |
|---|---|---|---|---|---|---|

## Underpinning

**Excavate preliminary trench by hand, maximum depth not exceeding**

| | | | | | | |
|---|---|---|---|---|---|---|
| 1.00m | m3 | 3.70 | 51.80 | 0.00 | 7.77 | 59.57 |
| 1.50m | m3 | 4.15 | 58.10 | 0.00 | 8.72 | 66.82 |
| 2.00m | m3 | 4.40 | 61.60 | 0.00 | 9.24 | 70.84 |
| 2.50m | m3 | 6.10 | 85.40 | 0.00 | 12.81 | 98.21 |
| 3.00m | m3 | 7.25 | 101.50 | 0.00 | 15.23 | 116.73 |

**Excavate trench by hand below existing foundations, maximum not exceeding**

| | | | | | | |
|---|---|---|---|---|---|---|
| 1.00m | m3 | 4.25 | 59.50 | 0.00 | 8.93 | 68.43 |
| 1.50m | m3 | 4.75 | 66.50 | 0.00 | 9.98 | 76.48 |
| 2.00m | m3 | 5.45 | 76.30 | 0.00 | 11.45 | 87.75 |
| 2.50m | m3 | 7.30 | 102.20 | 0.00 | 15.33 | 117.53 |
| 3.00m | m3 | 8.35 | 116.90 | 0.00 | 17.54 | 134.44 |

**Excavate and backfill working space, maximum depth not exceeding**

| | | | | | | |
|---|---|---|---|---|---|---|
| 1.00m | m3 | 5.00 | 70.00 | 0.00 | 10.50 | 80.50 |
| 1.50m | m3 | 5.25 | 73.50 | 0.00 | 11.03 | 84.53 |
| 2.00m | m3 | 5.90 | 82.60 | 0.00 | 12.39 | 94.99 |
| 2.50m | m3 | 8.10 | 113.40 | 0.00 | 17.01 | 130.41 |
| 3.00m | m3 | 9.25 | 129.50 | 0.00 | 19.43 | 148.93 |

**Open-boarded earthwork support to sides of preliminary trenches, distance between faces not exceeding 2.0m**

maximum depth not exceeding

| | | | | | | |
|---|---|---|---|---|---|---|
| 1.0m | m2 | 0.30 | 4.20 | 1.98 | 0.93 | 7.11 |

| | Unit | Labour hours | Labour cost £ | Materials £ | O & P £ | Total £ |
|---|---|---|---|---|---|---|
| maximum depth not exceeding 2.0m | m2 | 0.40 | 5.60 | 1.98 | 1.14 | 8.72 |

Close-boarded earthwork support to sides of preliminary trenches, distance between faces not exceeding 2.0m

| | Unit | Labour hours | Labour cost £ | Materials £ | O & P £ | Total £ |
|---|---|---|---|---|---|---|
| maximum depth not exceeding 1.0m | m2 | 0.85 | 11.90 | 3.39 | 2.29 | 17.58 |
| maximum depth not exceeding 2.0m | m2 | 1.10 | 15.40 | 3.39 | 2.82 | 21.61 |

Open-boarded earthwork support to sides of excavation trenches, distance between faces not exceeding 2.0m

| | Unit | Labour hours | Labour cost £ | Materials £ | O & P £ | Total £ |
|---|---|---|---|---|---|---|
| maximum depth not exceeding 1.0m | m2 | 0.35 | 4.90 | 1.98 | 1.03 | 7.91 |
| maximum depth not exceeding 2.0m | m2 | 0.45 | 6.30 | 1.98 | 1.24 | 9.52 |

Close-boarded earthwork support to sides of excavation trenches, distance between faces not exceeding 2.0m

| | Unit | Labour hours | Labour cost £ | Materials £ | O & P £ | Total £ |
|---|---|---|---|---|---|---|
| maximum depth not exceeding 1.0m | m2 | 0.95 | 13.30 | 3.39 | 2.50 | 19.19 |
| maximum depth not exceeding 2.0m | m2 | 1.20 | 16.80 | 3.39 | 3.03 | 23.22 |

| | Unit | Labour hours | Labour cost £ | Materials £ | O & P £ | Total £ |
|---|---|---|---|---|---|---|
| Cut away projecting concrete foundations, size | | | | | | |
| 150 × 150mm | m | 0.45 | 6.30 | 0.00 | 0.95 | 7.25 |
| 150 × 225mm | m | 0.55 | 7.70 | 0.00 | 1.16 | 8.86 |
| 150 × 300mm | m | 0.65 | 9.10 | 0.00 | 1.37 | 10.47 |
| Cut away projecting brickwork in footings, one brick thick | | | | | | |
| one course | m | 1.10 | 15.40 | 0.00 | 2.31 | 17.71 |
| two courses | m | 1.90 | 26.60 | 0.00 | 3.99 | 30.59 |
| three courses | m | 2.60 | 36.40 | 0.00 | 5.46 | 41.86 |
| Prepare underside of existing foundations to receive the new work, width | | | | | | |
| 300mm | m | 0.75 | 10.50 | 0.00 | 1.58 | 12.08 |
| 500mm | m | 1.15 | 16.10 | 0.00 | 2.42 | 18.52 |
| 750mm | m | 1.40 | 19.60 | 0.00 | 2.94 | 22.54 |
| 1000mm | m | 1.60 | 22.40 | 0.00 | 3.36 | 25.76 |
| 1200mm | m | 1.80 | 25.20 | 0.00 | 3.78 | 28.98 |
| Load surplus excavated material into barrows, wheel and deposit in temporary spoil heaps, average distance | | | | | | |
| 15m | m3 | 1.25 | 17.50 | 0.00 | 2.63 | 20.13 |
| 25m | m3 | 1.45 | 20.30 | 0.00 | 3.05 | 23.35 |
| 50m | m3 | 1.70 | 23.80 | 0.00 | 3.57 | 27.37 |
| Load surplus excavated material into barrows, wheel and deposit in temporary spoil heaps, average distance | | | | | | |
| 25m | m3 | 1.45 | 20.30 | 0.00 | 3.05 | 23.35 |
| 50m | m3 | 1.70 | 23.80 | 0.00 | 3.57 | 27.37 |

| | Unit | Labour hours | Labour cost £ | Materials £ | O & P £ | Total £ |
|---|---|---|---|---|---|---|
| Load surplus excavated material into barrows, wheel and deposit in skips or lorries, average distance | | | | | | |
| 25m | m3 | 1.40 | 19.60 | 0.00 | 2.94 | 22.54 |
| 50m | m3 | 1.65 | 23.10 | 0.00 | 3.47 | 26.57 |
| Level and compact bottom of excavation | m2 | 0.15 | 2.10 | 0.00 | 0.32 | 2.42 |
| Site-mixed concrete 1:3:6 40mm aggregate in foundations to underpinning, thickness | | | | | | |
| 150 to 300mm | m3 | 4.35 | 60.90 | 85.62 | 21.98 | 168.50 |
| 300 to 450mm | m3 | 4.05 | 56.70 | 85.62 | 21.35 | 163.67 |
| over 450mm | m3 | 3.65 | 51.10 | 85.62 | 20.51 | 157.23 |
| Site-mixed concrete 1:2:4 20mm aggregate in foundations to underpinning, thickness | | | | | | |
| 150 to 300mm | m3 | 4.35 | 60.90 | 91.44 | 22.85 | 175.19 |
| 300 to 450mm | m3 | 4.05 | 56.70 | 91.44 | 22.22 | 170.36 |
| over 450mm | m3 | 3.65 | 51.10 | 91.44 | 21.38 | 163.92 |
| Plain vertical formwork to sides of underpinned foundations, height | | | | | | |
| over 1m | m2 | 2.30 | 32.20 | 9.87 | 6.31 | 48.38 |
| not exceeding 250mm | m | 0.75 | 10.50 | 3.46 | 2.09 | 16.05 |
| 250 to 500mm | m | 1.25 | 17.50 | 5.98 | 3.52 | 27.00 |
| 500mm to 1m | m | 1.70 | 23.80 | 9.87 | 5.05 | 38.72 |

|  | Unit | Labour hours | Labour cost £ | Materials £ | O & P £ | Total £ |
|---|---|---|---|---|---|---|
| **Plain vertical formwork to sides of underpinned foundations, left in, height** | | | | | | |
| over 1m | m2 | 2.15 | 30.10 | 2.05 | 4.82 | 36.97 |
| not exceeding 250mm | m | 0.65 | 9.10 | 2.05 | 1.67 | 12.82 |
| 250 to 500mm | m | 1.15 | 16.10 | 2.05 | 2.72 | 20.87 |
| 500mm to 1m | m | 1.60 | 22.40 | 2.05 | 3.67 | 28.12 |
| **High yield deformed steel reinforcement bars, straight or bent** | | | | | | |
| 10mm diameter | m | 0.04 | 0.56 | 0.36 | 0.14 | 1.06 |
| 12mm diameter | m | 0.05 | 0.70 | 0.43 | 0.17 | 1.30 |
| 16mm diameter | m | 0.06 | 0.84 | 0.52 | 0.20 | 1.56 |
| 20mm diameter | m | 0.07 | 0.98 | 1.2 | 0.33 | 2.51 |
| 25mm diameter | m | 0.08 | 1.12 | 1.85 | 0.45 | 3.42 |
| **Common bricks basic price £200 per thousand in cement mortar in underpinning** | | | | | | |
| one brick thick | m2 | 5.20 | 119.60 | 2.05 | 18.25 | 139.90 |
| one and a half brick thick | m2 | 7.40 | 170.20 | 2.05 | 25.84 | 198.09 |
| two brick thick | m2 | 9.35 | 215.05 | 2.05 | 32.57 | 249.67 |
| **Class A engineering bricks in basic price £350 per thousand in cement mortar in underpinning** | | | | | | |
| one brick thick | m2 | 5.40 | 124.20 | 2.05 | 18.94 | 145.19 |
| one and a half brick thick | m2 | 7.60 | 174.80 | 2.05 | 26.53 | 203.38 |
| two brick thick | m2 | 9.50 | 218.50 | 2.05 | 33.08 | 253.63 |
| **Hessian-based bitumen damp proof course, bedded in cement mortar, horizontal** | | | | | | |
| over 225mm wide | m2 | 0.35 | 8.05 | 10.02 | 2.71 | 20.78 |
| 112mm wide | m | 0.05 | 1.15 | 1.43 | 0.39 | 2.97 |

| | Unit | Labour hours | Labour cost £ | Materials £ | O & P £ | Total £ |
|---|---|---|---|---|---|---|
| **Two courses of slates bedded in cement mortar, horizontal** | | | | | | |
| over 225mm wide | m2 | 2.90 | 66.70 | 35.95 | 15.40 | 118.05 |
| 225mm wide | m | 0.85 | 19.55 | 3.47 | 3.45 | 26.47 |
| **Wedge and pin new work to soffit of existing with slates in cement mortar, width** | | | | | | |
| one brick thick | m | 1.90 | 43.70 | 7.48 | 7.68 | 58.86 |
| one and a half brick thick | m | 2.40 | 55.20 | 10.42 | 9.84 | 75.46 |
| two brick thick | m | 2.80 | 64.40 | 15.37 | 11.97 | 91.74 |

## Temporary screens

| | Unit | Labour hours | Labour cost £ | Materials £ | O & P £ | Total £ |
|---|---|---|---|---|---|---|
| **Erect, maintain and remove temporary screens consisting of 50 × 50mm softwood framing covered one side with** | | | | | | |
| hardboard (three uses) | m2 | 1.15 | 18.40 | 1.57 | 3.00 | 22.97 |
| insulation board (three uses) | m2 | 1.15 | 18.40 | 2.23 | 3.09 | 23.72 |
| polythene sheeting (three uses) | m2 | 0.80 | 12.80 | 0.15 | 1.94 | 14.89 |

| | Unit | Labour hours | Labour cost £ | Materials £ | O & P £ | Total £ |
|---|---|---|---|---|---|---|

## SPOT ITEMS

### Brickwork, blockwork and masonry

Take out existing fireplace, fill opening with 100mm thick block-work plastered one side, fix new skirting to match existing, make good flooring where hearth removed

| | Unit | Labour hours | Labour cost £ | Materials £ | O & P £ | Total £ |
|---|---|---|---|---|---|---|
| medium size fireplace | nr | 12.00 | 192.00 | 17.78 | 31.47 | 241.25 |
| large size fireplace | nr | 14.00 | 224.00 | 28.58 | 37.89 | 290.47 |

Cut out existing projecting chimney breast, size 1350 × 350mm, including making good flooring and ceiling to match existing and plastering wall where breast removed

| | Unit | Labour hours | Labour cost £ | Materials £ | O & P £ | Total £ |
|---|---|---|---|---|---|---|
| one storey | nr | 36.00 | 576.00 | 45.47 | 93.22 | 714.69 |

Take down chimney stack to 100mm below roof level, seal flue with slates in cement mortar, make good roof timbers

| | Unit | Labour hours | Labour cost £ | Materials £ | O & P £ | Total £ |
|---|---|---|---|---|---|---|
| slated roof | nr | 46.00 | 736.00 | 115.44 | 127.72 | 979.16 |
| tiled roof | nr | 46.00 | 736.00 | 74.62 | 121.59 | 932.21 |

Cut out single brick in half brick wall and replace in gauged mortar

| | Unit | Labour hours | Labour cost £ | Materials £ | O & P £ | Total £ |
|---|---|---|---|---|---|---|
| commons | nr | 0.25 | 4.00 | 0.25 | 0.64 | 4.89 |
| facings | nr | 0.30 | 4.80 | 0.51 | 0.80 | 6.11 |

| | Unit | Labour hours | Labour cost £ | Materials £ | O & P £ | Total £ |
|---|---|---|---|---|---|---|
| Cut out decayed brickwork in half brick wall in areas 0.5 to 1m2 and replace in gauged mortar | | | | | | |
| commons | nr | 5.20 | 83.20 | 14.63 | 14.67 | 112.50 |
| facings | nr | 6.00 | 96.00 | 2.05 | 14.71 | 112.76 |
| Cut out decayed brickwork in one brick wall in areas 0.5 to 1m2 and replace in gauged mortar | | | | | | |
| commons | nr | 7.60 | 121.60 | 29.27 | 22.63 | 173.50 |
| facings | nr | 8.80 | 140.80 | 51.15 | 28.79 | 220.74 |
| Cut out vertical, horizontal or stepped cracks in half brick wall, replace average 350mm wide with new bricks in gauged mortar | | | | | | |
| commons | m | 2.75 | 44.00 | 5.45 | 7.42 | 56.87 |
| facings | m | 3.10 | 49.60 | 13.88 | 9.52 | 73.00 |
| Cut out vertical, horizontal or stepped cracks in one brick wall, replace average 350mm wide with new bricks in gauged mortar | | | | | | |
| commons | m | 5.20 | 83.20 | 10.93 | 14.12 | 108.25 |
| facings | m | 5.80 | 92.80 | 27.73 | 18.08 | 138.61 |
| Cut out defective brick-on-end soldier arch to half brick wall, and replace with new bricks in gauged mortar | | | | | | |
| commons | m | 3.00 | 48.00 | 8.29 | 8.44 | 64.73 |
| facings | m | 3.40 | 54.40 | 12.68 | 10.06 | 77.14 |

| | Unit | Labour hours | Labour cost £ | Materials £ | O & P £ | Total £ |
|---|---|---|---|---|---|---|
| Cut out defective brick-on-end soldier arch to one brick wall, and replace with new bricks in gauged mortar | | | | | | |
| commons | m | 3.80 | 60.80 | 16.59 | 11.61 | 89.00 |
| facings | m | 4.20 | 67.20 | 25.37 | 13.89 | 106.46 |
| Cut out defective terracotta air brick and replace | | | | | | |
| 215 × 65mm | nr | 0.35 | 5.60 | 2.56 | 1.22 | 9.38 |
| 215 × 140mm | nr | 0.50 | 8.00 | 3.44 | 1.72 | 13.16 |
| 215 × 215mm | nr | 0.65 | 10.40 | 9.54 | 2.99 | 22.93 |
| Rake out joints in gauged mortar, refix loose flashing and point up on completion | | | | | | |
| horizontal | m | 0.45 | 7.20 | 0.51 | 1.16 | 8.87 |
| stepped | m | 0.65 | 10.40 | 0.75 | 1.67 | 12.82 |
| Rake out joints and point up in gauged mortar | m2 | 0.65 | 10.40 | 0.75 | 1.67 | 12.82 |
| Cut out one course of half brick wall, insert hessian-based damp course 112mm wide and replace with new bricks in gauged mortar | | | | | | |
| commons | m | 2.00 | 32.00 | 3.33 | 5.30 | 40.63 |
| facings | m | 2.00 | 32.00 | 3.78 | 5.37 | 41.15 |
| Cut out one course of one brick wall, insert hessian-based damp course 112mm wide and replace with new bricks in gauged mortar | | | | | | |
| commons | m | 3.25 | 52.00 | 6.68 | 8.80 | 67.48 |
| facings | m | 3.25 | 52.00 | 7.59 | 8.94 | 68.53 |

| | Unit | Labour hours | Labour cost £ | Materials £ | O & P £ | Total £ |
|---|---|---|---|---|---|---|
| **Cut out single block, replace with new in gauged mortar, thickness** | | | | | | |
| 100mm | nr | 0.30 | 4.80 | 1.44 | 0.94 | 7.18 |
| 140mm | nr | 0.40 | 6.40 | 2.29 | 1.30 | 9.99 |
| 190mm | nr | 0.50 | 8.00 | 2.89 | 1.63 | 12.52 |
| 255mm | nr | 0.60 | 9.60 | 3.91 | 2.03 | 15.54 |
| **Rake out joints of random rubble walling and point up in gauged mortar** | | | | | | |
| flush pointing | m2 | 0.50 | 8.00 | 0.64 | 1.30 | 9.94 |
| weather pointing | m2 | 0.55 | 8.80 | 0.64 | 1.42 | 10.86 |

## Roofing

| | Unit | Labour hours | Labour cost £ | Materials £ | O & P £ | Total £ |
|---|---|---|---|---|---|---|
| **Take up roof coverings from pitched roof** | | | | | | |
| tiles | m2 | 0.80 | 12.80 | 0.00 | 1.92 | 14.72 |
| slates | m2 | 0.80 | 12.80 | 0.00 | 1.92 | 14.72 |
| timber boarding | m2 | 1.00 | 16.00 | 0.00 | 2.40 | 18.40 |
| metal sheeting | m2 | 0.20 | 3.20 | 0.00 | 0.48 | 3.68 |
| flat sheeting | m2 | 0.30 | 4.80 | 0.00 | 0.72 | 5.52 |
| corrugated sheeting | m2 | 0.30 | 4.80 | 0.00 | 0.72 | 5.52 |
| underfelt | m2 | 0.10 | 1.60 | 0.00 | 0.24 | 1.84 |
| **Take up roof coverings from flat roof** | | | | | | |
| bituminous felt | m2 | 0.25 | 4.00 | 0.00 | 0.60 | 4.60 |
| metal sheeting | m2 | 0.30 | 4.80 | 0.00 | 0.72 | 5.52 |
| woodwool slabs | m2 | 0.50 | 8.00 | 0.00 | 1.20 | 9.20 |
| firrings | m2 | 0.20 | 3.20 | 0.00 | 0.48 | 3.68 |

| | Unit | Labour hours | Labour cost £ | Materials £ | O & P £ | Total £ |
|---|---|---|---|---|---|---|
| **Take up roof coverings from pitched roof, carefully lay aside for reuse** | | | | | | |
| tiles | m2 | 1.10 | 17.60 | 0.00 | 2.64 | 20.24 |
| slates | m2 | 1.10 | 17.60 | 0.00 | 2.64 | 20.24 |
| metal sheeting | m2 | 0.50 | 8.00 | 0.00 | 1.20 | 9.20 |
| flat sheeting | m2 | 0.60 | 9.60 | 0.00 | 1.44 | 11.04 |
| corrugated sheeting | m2 | 0.60 | 9.60 | 0.00 | 1.44 | 11.04 |
| **Take up roof coverings from flat roof, carefully lay aside for reuse** | | | | | | |
| metal sheeting | m2 | 0.60 | 9.60 | 0.00 | 1.44 | 11.04 |
| woodwool slabs | m2 | 0.80 | 12.80 | 0.00 | 1.92 | 14.72 |
| **Inspect roof battens, refix loose and replace with new, size 38 × 25mm** | | | | | | |
| **25% of area** | | | | | | |
| 250mm centres | m2 | 0.14 | 2.24 | 0.38 | 0.39 | 3.01 |
| 450mm centres | m2 | 0.12 | 1.92 | 0.31 | 0.33 | 2.56 |
| 600mm centres | m2 | 0.10 | 1.60 | 0.25 | 0.28 | 2.13 |
| **50% of area** | | | | | | |
| 250mm centres | m2 | 0.26 | 4.16 | 0.78 | 0.74 | 5.68 |
| 450mm centres | m2 | 0.16 | 2.56 | 0.64 | 0.48 | 3.68 |
| 600mm centres | m2 | 0.18 | 2.88 | 0.51 | 0.51 | 3.90 |
| **75% of area** | | | | | | |
| 250mm centres | m2 | 0.36 | 5.76 | 1.16 | 1.04 | 7.96 |
| 450mm centres | m2 | 0.26 | 4.16 | 0.97 | 0.77 | 5.90 |
| 600mm centres | m2 | 0.22 | 3.52 | 0.78 | 0.65 | 4.95 |
| **100% of area** | | | | | | |
| 250mm centres | m2 | 0.44 | 7.04 | 2.01 | 1.36 | 10.41 |
| 450mm centres | m2 | 0.32 | 5.12 | 1.30 | 0.96 | 7.38 |
| 600mm centres | m2 | 0.38 | 6.08 | 1.04 | 1.07 | 8.19 |

| | Unit | Labour hours | Labour cost £ | Materials £ | O & P £ | Total £ |
|---|---|---|---|---|---|---|
| Remove single slipped slate and refix | nr | 1.00 | 16.00 | 1.04 | 2.56 | 19.60 |
| Remove single broken slate, renew with new Welsh blue slate | | | | | | |
| 405 × 255mm | nr | 1.20 | 19.20 | 1.57 | 3.12 | 23.89 |
| 510 × 255mm | nr | 1.20 | 19.20 | 3.02 | 3.33 | 25.55 |
| 610 × 305mm | nr | 1.20 | 19.20 | 6.39 | 3.84 | 29.43 |
| Remove slates in area approximately 1m2 and replace with Welsh blue slates previously laid aside | nr | 1.60 | 25.60 | 0.00 | 3.84 | 29.44 |
| Remove slates in area approximately 1m2 and replace with Welsh blue slates previously laid aside | | | | | | |
| 405 × 255mm | nr | 1.80 | 28.80 | 0.00 | 4.32 | 33.12 |
| 510 × 255mm | nr | 1.70 | 27.20 | 0.00 | 4.08 | 31.28 |
| 610 × 305mm | nr | 1.60 | 25.60 | 0.00 | 3.84 | 29.44 |
| Remove double course at eaves and fix new Welsh blue slates | | | | | | |
| 405 × 255mm | m | 0.70 | 11.20 | 56.85 | 10.21 | 78.26 |
| 510 × 255mm | m | 0.70 | 11.20 | 58.41 | 10.44 | 80.05 |
| 610 × 305mm | m | 0.70 | 11.20 | 64.20 | 11.31 | 86.71 |
| Remove single verge undercloak course and renew | | | | | | |
| 405 × 255mm | m | 0.90 | 14.40 | 21.84 | 5.44 | 41.68 |
| 510 × 255mm | m | 0.90 | 14.40 | 23.17 | 5.64 | 43.21 |
| 610 × 305mm | m | 0.90 | 14.40 | 25.39 | 5.97 | 45.76 |
| Remove single slipped tile and refix | nr | 0.30 | 4.80 | 0.00 | 0.72 | 5.52 |

| | Unit | Labour hours | Labour cost £ | Materials £ | O & P £ | Total £ |
|---|---|---|---|---|---|---|
| **Remove single broken tile and renew** | | | | | | |
| Marley plain tile | nr | 0.30 | 4.80 | 0.62 | 0.81 | 6.23 |
| Marley Ludlow Plus tile | nr | 0.30 | 4.80 | 0.93 | 0.86 | 6.59 |
| Marley Modern tile | nr | 0.30 | 4.80 | 1.44 | 0.94 | 7.18 |
| Redland Renown tile | nr | 0.30 | 4.80 | 1.44 | 0.94 | 7.18 |
| Redland Norfolk tile | nr | 0.30 | 4.80 | 1.44 | 0.94 | 7.18 |
| **Remove tiles in area approximately 1m2 and replace with tiles previously laid aside** | | | | | | |
| Marley Plain tile | nr | 1.80 | 28.80 | 35.45 | 9.64 | 73.89 |
| Marley Ludlow Plus tile | nr | 1.20 | 19.20 | 12.34 | 4.73 | 36.27 |
| Marley Modern tile | nr | 1.10 | 17.60 | 19.50 | 5.57 | 42.67 |
| Lafarge Renown tile | nr | 1.10 | 17.60 | 18.50 | 5.42 | 41.52 |
| Lafarge Norfolk tile | nr | 1.15 | 18.40 | 14.50 | 4.94 | 37.84 |
| **Take off defective ridge capping and refix including pointing in mortar** | m | 1.10 | 17.60 | 1.80 | 2.91 | 22.31 |

**Carpentry and joinery**

Take down, cut out or demolish structural timbers and load into skips

| | Unit | Labour hours | Labour cost £ | Materials £ | O & P £ | Total £ |
|---|---|---|---|---|---|---|
| structural timbers | | | | | | |
| 50 × 100mm | m | 0.10 | 1.60 | 0.00 | 0.24 | 1.84 |
| 50 × 150mm | m | 0.12 | 1.92 | 0.00 | 0.29 | 2.21 |
| 75 × 100mm | m | 0.14 | 2.24 | 0.00 | 0.34 | 2.58 |
| 75 × 150mm | m | 0.16 | 2.56 | 0.00 | 0.38 | 2.94 |
| 100 × 150mm | m | 0.18 | 2.88 | 0.00 | 0.43 | 3.31 |
| 100 × 200mm | m | 0.20 | 3.20 | 0.00 | 0.48 | 3.68 |
| roof boarding | m2 | 0.28 | 4.48 | 0.00 | 0.67 | 5.15 |
| floor boarding | m2 | 0.22 | 3.52 | 0.00 | 0.53 | 4.05 |

| | Unit | Labour hours | Labour cost £ | Materials £ | O & P £ | Total £ |
|---|---|---|---|---|---|---|
| stud partition plasterboard | | | | | | |
| both sides | m2 | 0.75 | 12.00 | 0.00 | 1.80 | 13.80 |
| skirtings and grounds | | | 0.00 | 0.00 | 0.00 | 0.00 |
| 100mm high | m | 0.08 | 1.28 | 0.00 | 0.19 | 1.47 |
| 150mm high | m | 0.09 | 1.44 | 0.00 | 0.22 | 1.66 |
| 200mm high | m | 0.10 | 1.60 | 0.00 | 0.24 | 1.84 |
| rails | | | | | | |
| 50mm high | m | 0.05 | 0.80 | 0.00 | 0.12 | 0.92 |
| 75mm high | m | 0.06 | 0.96 | 0.00 | 0.14 | 1.10 |
| 100mm high | m | 0.07 | 1.12 | 0.00 | 0.17 | 1.29 |
| fittings | | | | | | |
| wall cupboards | nr | 0.25 | 4.00 | 0.00 | 0.60 | 4.60 |
| floor units | nr | 0.20 | 3.20 | 0.00 | 0.48 | 3.68 |
| sink units | nr | 0.25 | 4.00 | 0.00 | 0.60 | 4.60 |
| staircase, 900mm wide | | | | | | |
| straight flight | nr | 4.00 | 64.00 | 0.00 | 9.60 | 73.60 |
| landing | nr | 1.50 | 24.00 | 0.00 | 3.60 | 27.60 |
| doors, frames and linings | | | | | | |
| single, internal | nr | 0.40 | 6.40 | 0.00 | 0.96 | 7.36 |
| single, external | nr | 0.60 | 9.60 | 0.00 | 1.44 | 11.04 |
| double, internal | nr | 0.60 | 9.60 | 0.00 | 1.44 | 11.04 |
| double, external | nr | 0.80 | 12.80 | 0.00 | 1.92 | 14.72 |
| windows | | | | | | |
| casement, 1200 × 900mm | nr | 0.50 | 8.00 | 0.00 | 1.20 | 9.20 |
| casement, 1800 × 900mm | nr | 0.60 | 9.60 | 0.00 | 1.44 | 11.04 |
| sash, 900 × 1500mm | nr | 0.80 | 12.80 | 0.00 | 1.92 | 14.72 |
| sash, 1800 × 900mm | nr | 0.90 | 14.40 | 0.00 | 2.16 | 16.56 |
| ironmongery | | | | | | |
| bolt | nr | 0.20 | 3.20 | 0.00 | 0.48 | 3.68 |
| deadlock | nr | 0.25 | 4.00 | 0.00 | 0.60 | 4.60 |
| mortice lock | nr | 0.35 | 5.60 | 0.00 | 0.84 | 6.44 |
| mortice latch | nr | 0.35 | 5.60 | 0.00 | 0.84 | 6.44 |
| cylinder lock | nr | 0.25 | 4.00 | 0.00 | 0.60 | 4.60 |
| door closer | nr | 0.35 | 5.60 | 0.00 | 0.84 | 6.44 |
| casement stay | nr | 0.15 | 2.40 | 0.00 | 0.36 | 2.76 |
| casement fastener | nr | 0.15 | 2.40 | 0.00 | 0.36 | 2.76 |
| toilet roll holder | nr | 0.15 | 2.40 | 0.00 | 0.36 | 2.76 |
| shelf bracket | nr | 0.15 | 2.40 | 0.00 | 0.36 | 2.76 |

| | Unit | Labour hours | Labour cost £ | Materials £ | O & P £ | Total £ |
|---|---|---|---|---|---|---|
| Cut out defective joists or rafters and replace with new | | | | | | |
| 50 × 75mm | nr | 0.35 | 5.60 | 1.51 | 1.07 | 8.18 |
| 50 × 100mm | nr | 0.40 | 6.40 | 1.57 | 1.20 | 9.17 |
| 50 × 150mm | nr | 0.60 | 9.60 | 2.35 | 1.79 | 13.74 |
| 75 × 100mm | nr | 0.65 | 10.40 | 3.17 | 2.04 | 15.61 |
| 75 × 150mm | nr | 0.75 | 12.00 | 3.87 | 2.38 | 18.25 |
| Cut out defective skirting and renew | | | | | | |
| 75mm high | m | 0.30 | 4.80 | 2.48 | 1.09 | 8.37 |
| 100mm high | m | 0.35 | 5.60 | 2.91 | 1.28 | 9.79 |
| 150mm high | m | 0.40 | 6.40 | 3.77 | 1.53 | 11.70 |
| Ease, adjust and oil | | | | | | |
| door | nr | 0.60 | 9.60 | 0.00 | 1.44 | 11.04 |
| casement window | nr | 0.40 | 6.40 | 0.00 | 0.96 | 7.36 |
| sash window including renewing cords | nr | 1.50 | 24.00 | 5.48 | 4.42 | 33.90 |
| Take up defective flooring and replace with 25mm thick plain edged boarding | | | | | | |
| areas less than 1m2 | m2 | 1.50 | 24.00 | 13.78 | 5.67 | 43.45 |
| areas more than 1m2 | m2 | 1.20 | 19.20 | 13.78 | 4.95 | 37.93 |
| Refix loose floorboards including punching in protruding nails | m2 | 0.20 | 3.20 | 0.00 | 0.48 | 3.68 |
| Take down existing door, lay aside, piece up frame or lining where butts removed, rehang door to opposite hand on existing butts and hardware | | | | | | |
| single, internal | nr | 1.50 | 24.00 | 2.02 | 3.90 | 29.92 |
| single, external | nr | 1.70 | 27.20 | 2.02 | 4.38 | 33.60 |

| | Unit | Labour hours | Labour cost £ | Materials £ | O & P £ | Total £ |
|---|---|---|---|---|---|---|

**Finishings**

Take down plasterboard sheeting from

| | | | | | | |
|---|---|---|---|---|---|---|
| studded walls | m2 | 0.35 | 5.60 | 0.00 | 0.84 | 6.44 |
| ceilings | m2 | 0.40 | 6.40 | 0.00 | 0.96 | 7.36 |

Hack off plaster from

| | | | | | | |
|---|---|---|---|---|---|---|
| walls | m2 | 0.25 | 4.00 | 0.00 | 0.60 | 4.60 |
| ceiling | m2 | 0.30 | 4.80 | 0.00 | 0.72 | 5.52 |

Make good plaster to walls where wall removed

| | | | | | | |
|---|---|---|---|---|---|---|
| 100mm wide | m2 | 0.95 | 15.20 | 1.96 | 2.57 | 19.73 |
| 150mm wide | m2 | 1.05 | 16.80 | 2.10 | 2.84 | 21.74 |
| 200mm wide | m2 | 1.15 | 18.40 | 2.23 | 3.09 | 23.72 |

Make good plaster to ceilings where wall removed

| | | | | | | |
|---|---|---|---|---|---|---|
| 100mm wide | m2 | 1.10 | 17.60 | 1.96 | 2.93 | 22.49 |
| 150mm wide | m2 | 1.15 | 18.40 | 2.10 | 3.08 | 23.58 |
| 200mm wide | m2 | 1.20 | 19.20 | 2.23 | 3.21 | 24.64 |

Cut out cracks in plasterwork and make good

| | | | | | | |
|---|---|---|---|---|---|---|
| walls | m2 | 0.35 | 5.60 | 0.31 | 0.89 | 6.80 |
| ceiling | m2 | 0.40 | 6.40 | 0.31 | 1.01 | 7.72 |

Hack off wall tiling and make good surface

| | | | | | | |
|---|---|---|---|---|---|---|
| good surface | m2 | 0.75 | 12.00 | 6.00 | 2.70 | 20.70 |

| | Unit | Labour hours | Labour cost £ | Materials £ | O & P £ | Total £ |
|---|---|---|---|---|---|---|
| **Plumbing** | | | | | | |
| Remove sanitary fittings and supports, seal off supply and waste pipes and prepare to receive new fittings | | | | | | |
| bath | nr | 3.50 | 61.25 | 13.17 | 11.16 | 85.58 |
| sink | nr | 3.50 | 61.25 | 13.17 | 11.16 | 85.58 |
| lavatory basin | nr | 3.25 | 56.88 | 7.90 | 9.72 | 74.49 |
| WC | nr | 3.75 | 65.63 | 7.90 | 11.03 | 84.55 |
| shower cubicle | nr | 4.00 | 70.00 | 10.54 | 12.08 | 92.62 |
| bidet | nr | 3.25 | 56.88 | 7.90 | 9.72 | 74.49 |
| Take down length of existing gutter, prepare ends and install new length of gutter to existing brackets | | | | | | |
| PVC-U | | | | | | |
| 76mm | nr | 0.60 | 10.50 | 9.36 | 2.98 | 22.84 |
| 112mm | nr | 0.65 | 11.38 | 11.46 | 3.43 | 26.26 |
| cast iron | | | | | | |
| 100mm | nr | 1.12 | 19.60 | 28.36 | 7.19 | 55.15 |
| Take down existing brackets from fascias and replace with galvanised steel repair brackets at 1m centres | m | 0.30 | 5.25 | 2.82 | 1.21 | 9.28 |
| Take down length of existing pipe, prepare ends and install new length of pipe | | | | | | |
| PVC-U | | | | | | |
| 68mm diameter | nr | 0.75 | 13.13 | 10.36 | 3.52 | 27.01 |
| 68mm square | nr | 0.75 | 13.13 | 11.38 | 3.68 | 28.18 |
| cast iron | | | | | | |
| 75mm | nr | 0.90 | 15.75 | 41.94 | 8.65 | 66.34 |
| 100mm | nr | 1.05 | 18.38 | 57.84 | 11.43 | 87.65 |

| | Unit | Labour hours | Labour cost £ | Materials £ | O & P £ | Total £ |
|---|---|---|---|---|---|---|
| **Take down fittings and install new** | | | | | | |
| PVC-U | | | | | | |
| 68mm diameter bend | nr | 0.50 | 8.75 | 3.87 | 1.89 | 14.51 |
| 68mm offset | nr | 0.50 | 8.75 | 9.02 | 2.67 | 20.44 |
| 68mm branch | nr | 0.50 | 8.75 | 9.22 | 2.70 | 20.67 |
| 68mm shoe | nr | 0.75 | 13.13 | 8.33 | 3.22 | 24.67 |
| Cast iron | | | | | | |
| 75mm diameter bend | nr | 0.60 | 10.50 | 7.76 | 2.74 | 21.00 |
| 75mm offset | nr | 0.60 | 10.50 | 19.61 | 4.52 | 34.63 |
| 75mm branch | nr | 0.60 | 10.50 | 8.52 | 2.85 | 21.87 |
| 75mm shoe | nr | 0.60 | 10.50 | 18.39 | 4.33 | 33.22 |
| Cast iron | | | | | | |
| 100mm diameter bend | nr | 0.70 | 12.25 | 26.13 | 5.76 | 44.14 |
| 100mm offset | nr | 0.70 | 12.25 | 36.16 | 7.26 | 55.67 |
| 100mm branch | nr | 0.70 | 12.25 | 10.43 | 3.40 | 26.08 |
| 100mm shoe | nr | 0.70 | 12.25 | 22.01 | 5.14 | 39.40 |
| **Cut out 500mm length of copper pipe, install new pipe with compression fittings at each end** | | | | | | |
| 15mm | nr | 0.80 | 14.00 | 5.10 | 2.87 | 21.97 |
| 22mm | nr | 0.90 | 15.75 | 8.95 | 3.71 | 28.41 |
| **Take off existing radiator valve and replace with new** | nr | 0.90 | 15.75 | 11.67 | 4.11 | 31.53 |

| | Unit | Labour hours | Labour cost £ | Materials £ | O & P £ | Total £ |
|---|---|---|---|---|---|---|

Take out existing galvanised steel water storage tank and install new plastic tank complete with ball valve, lid and insulation including cutting holes, make up pipework and connectors to existing pipework

| | Unit | Labour hours | Labour cost £ | Materials £ | O & P £ | Total £ |
|---|---|---|---|---|---|---|
| SC10, 18 litres | nr | 3.60 | 63.00 | 59.28 | 18.34 | 140.62 |
| SC15, 36 litres | nr | 3.70 | 64.75 | 64.21 | 19.34 | 148.30 |
| SC20, 54 litres | nr | 3.75 | 65.63 | 71.02 | 20.50 | 157.14 |
| SC25, 68 litres | nr | 3.80 | 66.50 | 79.98 | 21.97 | 168.45 |
| SC30, 86 litres | nr | 4.90 | 85.75 | 86.40 | 25.82 | 197.97 |
| SC40, 114 litres | nr | 5.00 | 87.50 | 90.08 | 26.64 | 204.22 |

**Glazing**

| | Unit | Labour hours | Labour cost £ | Materials £ | O & P £ | Total £ |
|---|---|---|---|---|---|---|
| Hack out glass and remove | m2 | 0.45 | 7.20 | 0.00 | 1.08 | 8.28 |
| Clean rebates, remove sprigs or clips and prepare for reglazing | m | 0.20 | 3.20 | 0.00 | 0.48 | 3.68 |

**Painting**

Prepare, wash down painted surfaces, rub down to receive new paintwork

| | Unit | Labour hours | Labour cost £ | Materials £ | O & P £ | Total £ |
|---|---|---|---|---|---|---|
| brickwork | m2 | 0.14 | 2.24 | 0.00 | 0.34 | 2.58 |
| blockwork | m2 | 0.15 | 2.40 | 0.00 | 0.36 | 2.76 |
| plasterwork | m2 | 0.12 | 1.92 | 0.00 | 0.29 | 2.21 |

Prepare, wash down previously painted wood surfaces, rub down to receive new paintwork surfaces

| | Unit | Labour hours | Labour cost £ | Materials £ | O & P £ | Total £ |
|---|---|---|---|---|---|---|
| over 300mm girth | m2 | 0.28 | 4.48 | 0.00 | 0.67 | 5.15 |
| isolated surfaces not exceeding 300mm girth | m | 0.10 | 1.60 | 0.00 | 0.24 | 1.84 |
| isolated surfaces not exceeding 0.5m2 | nr | 0.12 | 1.92 | 0.00 | 0.29 | 2.21 |

| | Unit | Labour hours | Labour cost £ | Materials £ | O & P £ | Total £ |
|---|---|---|---|---|---|---|

## Wallpapering

Strip off one layer of existing
paper, stop cracks and rub down
to receive new paper

| | Unit | Labour hours | Labour cost £ | Materials £ | O & P £ | Total £ |
|---|---|---|---|---|---|---|
| **woodchip** | | | | | | |
| walls | m2 | 0.18 | 2.88 | 0.00 | 0.43 | 3.31 |
| walls in staircase areas | m2 | 0.20 | 3.20 | 0.00 | 0.48 | 3.68 |
| ceilings | m2 | 0.22 | 3.52 | 0.00 | 0.53 | 4.05 |
| ceilings in staircase areas | m2 | 0.24 | 3.84 | 0.00 | 0.58 | 4.42 |
| **vinyl** | | | | | | |
| walls | m2 | 0.23 | 3.68 | 0.00 | 0.55 | 4.23 |
| walls in staircase areas | m2 | 0.25 | 4.00 | 0.00 | 0.60 | 4.60 |
| ceilings | m2 | 0.27 | 4.32 | 0.00 | 0.65 | 4.97 |
| ceilings in staircase areas | m2 | 0.29 | 4.64 | 0.00 | 0.70 | 5.34 |
| **standard patterned** | | | | | | |
| walls | m2 | 0.21 | 3.36 | 0.00 | 0.50 | 3.86 |
| walls in staircase areas | m2 | 0.23 | 3.68 | 0.00 | 0.55 | 4.23 |
| ceilings | m2 | 0.25 | 4.00 | 0.00 | 0.60 | 4.60 |
| ceilings in staircase areas | m2 | 0.27 | 4.32 | 0.00 | 0.65 | 4.97 |

# Part Two

## DAMAGE REPAIRS

Emergency measures

Fire damage

Flood damage

Gale damage

Theft damage

| | Unit | Labour hours | Labour cost £ | Plant £ | O & P £ | Total £ |
|---|---|---|---|---|---|---|
| **EMERGENCY MEASURES** | | | | | | |
| **Flooding** | | | | | | |
| Install hired pump and hoses in position | Item | 1.00 | 12.00 | 0.00 | 1.80 | 13.80 |
| Hire diaphragm pump and hoses | | | | | | |
| 50mm | week | 1.00 | 0.00 | 45.00 | 6.75 | 51.75 |
| 75mm | week | 1.00 | 0.00 | 55.00 | 8.25 | 63.25 |

| | Unit | Labour hours | Labour cost £ | Materials £ | O & P £ | Total £ |
|---|---|---|---|---|---|---|
| **Hoardings and screens** | | | | | | |
| Temporary screens and hoardings 2m high consisting of 22mm thick exterior quality plywood fixed to 100 × 50mm posts and rails | m | 2.60 | 31.20 | 36.93 | 10.22 | 78.35 |
| Hire tarpaulin sheeting size 5x x 4m fixed in position | day | 0.00 | 0.00 | 4.51 | 0.68 | 5.19 |
| Plywood sheeting blocking up window or door opening | m2 | 1.00 | 12.00 | 18.47 | 4.57 | 35.04 |
| **Shoring** | | | | | | |
| Timber dead shores consisting of 200 × 200mm shores, 250 × 50mm plates and 200 × 50mm braces at centres of | | | | | | |
| 2m | m2 | 3.80 | 45.60 | 86.85 | 19.87 | 152.32 |
| 3m | m2 | 3.00 | 36.00 | 68.63 | 15.69 | 120.32 |
| 4m | m2 | 2.20 | 26.40 | 49.42 | 11.37 | 87.19 |

| | Unit | Labour hours | Labour cost £ | Plant £ | O & P £ | Total £ |
|---|---|---|---|---|---|---|
| Timber raking and flying shores consisting of 200 × 200mm shores, 250 × 50mm plates and 200 × 50mm braces to gable end of two-storey house | Item | 70.00 | 840.00 | 560.99 | 210.15 | 1611.14 |

**Access towers**

| | Unit | Labour hours | Labour cost £ | Plant £ | O & P £ | Total £ |
|---|---|---|---|---|---|---|
| Hire narrow width access tower size 0.85 × 1.8m, height | | | | | | |
| 7.20m | week | 1.00 | 0.00 | 175.00 | 26.25 | 201.25 |
| 6.20m | week | 1.00 | 0.00 | 157.00 | 23.55 | 180.55 |
| 5.20m | week | 1.00 | 0.00 | 138.50 | 20.78 | 159.28 |
| 4.20m | week | 1.00 | 0.00 | 120.00 | 18.00 | 138.00 |
| Hire double width access tower size 1.45 × 2.5m, height | | | | | | |
| 7.20m | week | 1.00 | 0.00 | 257.00 | 38.55 | 295.55 |
| 6.20m | week | 1.00 | 0.00 | 235.00 | 35.25 | 270.25 |
| 5.20m | week | 1.00 | 0.00 | 213.00 | 31.95 | 244.95 |
| 4.20m | week | 1.00 | 0.00 | 191.00 | 28.65 | 219.65 |

| | Unit | Labour hours | Labour cost £ | Materials £ | O & P £ | Total £ |
|---|---|---|---|---|---|---|

## FIRE DAMAGE

### Window replacement

Take out existing window, prepare
jambs and cill to receive new
PVC-U

| | Unit | Labour hours | Labour cost £ | Materials £ | O & P £ | Total £ |
|---|---|---|---|---|---|---|
| size 600 × 900mm | nr | 2.00 | 32.00 | 127.54 | 23.93 | 183.47 |
| size 600 × 1200mm | nr | 2.00 | 32.00 | 168.84 | 30.13 | 230.97 |
| size 1200 × 1200mm | nr | 2.50 | 40.00 | 194.85 | 35.23 | 270.08 |
| size 1800 × 1200mm | nr | 3.00 | 48.00 | 205.95 | 38.09 | 292.04 |

softwood painted

| | Unit | Labour hours | Labour cost £ | Materials £ | O & P £ | Total £ |
|---|---|---|---|---|---|---|
| size 630 × 900mm | nr | 2.00 | 32.00 | 127.22 | 23.88 | 183.10 |
| size 915 × 900mm | nr | 2.00 | 32.00 | 138.03 | 25.50 | 195.53 |
| size 915 × 1200mm | nr | 2.50 | 40.00 | 148.56 | 28.28 | 216.84 |
| size 1200 × 1200mm | nr | 3.00 | 48.00 | 197.46 | 36.82 | 282.28 |

hardwood stained

| | Unit | Labour hours | Labour cost £ | Materials £ | O & P £ | Total £ |
|---|---|---|---|---|---|---|
| size 915 × 1050mm | nr | 2.00 | 32.00 | 287.03 | 47.85 | 366.88 |
| size 915 × 1500mm | nr | 2.00 | 32.00 | 298.87 | 49.63 | 380.50 |
| size 1200 × 1500mm | nr | 2.50 | 40.00 | 341.01 | 57.15 | 438.16 |
| size 1770 × 1200mm | nr | 3.00 | 48.00 | 306.92 | 53.24 | 408.16 |

Take out existing bay window,
prepare jambs and cill to receive
new
PVC-U

| | Unit | Labour hours | Labour cost £ | Materials £ | O & P £ | Total £ |
|---|---|---|---|---|---|---|
| size 1800 × 900mm | nr | 3.00 | 48.00 | 250.56 | 44.78 | 343.34 |
| size 1800 × 1200mm | nr | 3.00 | 48.00 | 280.80 | 49.32 | 378.12 |
| size 2400 × 900mm | nr | 3.50 | 56.00 | 481.68 | 80.65 | 618.33 |
| size 2400 × 1200mm | nr | 3.50 | 56.00 | 515.16 | 85.67 | 656.83 |

softwood painted

| | Unit | Labour hours | Labour cost £ | Materials £ | O & P £ | Total £ |
|---|---|---|---|---|---|---|
| size 1800 × 900mm | nr | 3.00 | 48.00 | 178.20 | 33.93 | 260.13 |
| size 1800 × 1200mm | nr | 3.00 | 48.00 | 206.28 | 38.14 | 292.42 |
| size 2400 × 900mm | nr | 3.50 | 56.00 | 307.80 | 54.57 | 418.37 |
| size 2400 × 1200mm | nr | 3.50 | 56.00 | 335.88 | 58.78 | 450.66 |

| | Unit | Labour hours | Labour cost £ | Materials £ | O & P £ | Total £ |
|---|---|---|---|---|---|---|
| hardwood stained | | | | | | |
| size 1800 × 900mm | nr | 3.00 | 48.00 | 298.43 | 51.96 | 398.39 |
| size 1800 × 1200mm | nr | 3.00 | 48.00 | 334.23 | 57.33 | 439.56 |
| size 2400 × 900mm | nr | 3.50 | 56.00 | 380.99 | 65.55 | 502.54 |
| size 2400 × 1200mm | nr | 3.50 | 56.00 | 371.74 | 64.16 | 491.90 |

## Window repairs

Take off and replace defective
ironmongery to softwood windows

| | Unit | Labour hours | Labour cost £ | Materials £ | O & P £ | Total £ |
|---|---|---|---|---|---|---|
| casement fastener | nr | 0.20 | 3.20 | 5.62 | 1.32 | 10.14 |
| casement stay | nr | 0.20 | 3.20 | 6.90 | 1.52 | 11.62 |
| hinges | pair | 0.25 | 4.00 | 4.11 | 1.22 | 9.33 |
| cockspur fastener | nr | 0.20 | 3.20 | 11.46 | 2.20 | 16.86 |
| sash fastener | nr | 0.25 | 4.00 | 5.28 | 1.39 | 10.67 |

Take off and replace defective
ironmongery to hardwood windows

| | Unit | Labour hours | Labour cost £ | Materials £ | O & P £ | Total £ |
|---|---|---|---|---|---|---|
| casement fastener | nr | 0.30 | 4.80 | 4.54 | 1.40 | 10.74 |
| casement stay | nr | 0.30 | 4.80 | 6.90 | 1.76 | 13.46 |
| hinges | pair | 0.35 | 5.60 | 4.11 | 1.46 | 11.17 |
| cockspur fastener | nr | 0.30 | 4.80 | 11.48 | 2.44 | 18.72 |
| sash fastener | nr | 0.35 | 5.60 | 5.28 | 1.63 | 12.51 |

Take off and replace defective
ironmongery to PVC-U windows

| | Unit | Labour hours | Labour cost £ | Materials £ | O & P £ | Total £ |
|---|---|---|---|---|---|---|
| casement fastener | nr | 0.20 | 3.20 | 4.54 | 1.16 | 8.90 |
| casement stay | nr | 0.20 | 3.20 | 6.90 | 1.52 | 11.62 |
| hinges | pair | 0.25 | 4.00 | 4.11 | 1.22 | 9.33 |
| cockspur fastener | nr | 0.20 | 3.20 | 11.48 | 2.20 | 16.88 |
| sash fastener | nr | 0.25 | 4.00 | 5.28 | 1.39 | 10.67 |

Take out existing window cill and
replace

| | Unit | Labour hours | Labour cost £ | Materials £ | O & P £ | Total £ |
|---|---|---|---|---|---|---|
| softwood | m | 0.55 | 8.80 | 9.38 | 2.73 | 20.91 |
| hardwood | m | 0.65 | 10.40 | 27.84 | 5.74 | 43.98 |

| | Unit | Labour hours | Labour cost £ | Materials £ | O & P £ | Total £ |
|---|---|---|---|---|---|---|
| Take out existing window board and replace | | | | | | |
| softwood | m | 0.45 | 7.20 | 7.21 | 2.16 | 16.57 |
| hardwood | m | 0.60 | 9.60 | 23.24 | 4.93 | 37.77 |

## Door replacement

| | Unit | Labour hours | Labour cost £ | Materials £ | O & P £ | Total £ |
|---|---|---|---|---|---|---|
| Take off existing external door and frame and replace with new flush door | | | | | | |
| softwood | nr | 2.20 | 35.20 | 108.24 | 21.52 | 164.96 |
| hardwood | nr | 2.50 | 40.00 | 391.48 | 64.72 | 496.20 |
| PVC-U | nr | 2.20 | 35.20 | 360.40 | 59.34 | 454.94 |
| panelled door | | | | | | |
| softwood | nr | 2.20 | 35.20 | 137.16 | 25.85 | 198.21 |
| hardwood | nr | 2.50 | 40.00 | 429.21 | 70.38 | 539.59 |
| PVC-U | nr | 2.20 | 35.20 | 402.79 | 65.70 | 503.69 |
| half glazed door | | | | | | |
| softwood | nr | 2.20 | 35.20 | 137.16 | 25.85 | 198.21 |
| hardwood | nr | 2.50 | 40.00 | 429.21 | 70.38 | 539.59 |
| PVC-U | nr | 2.20 | 35.20 | 402.79 | 65.70 | 503.69 |
| fully glazed door | | | | | | |
| softwood | nr | 2.80 | 44.80 | 110.04 | 23.23 | 178.07 |
| hardwood | nr | 3.20 | 51.20 | 405.65 | 68.53 | 525.38 |
| PVC-U | nr | 2.80 | 44.80 | 374.75 | 62.93 | 482.48 |
| fully glazed patio doors | | | | | | |
| softwood | pair | 3.80 | 60.80 | 246.94 | 46.16 | 353.90 |
| hardwood | pair | 4.30 | 68.80 | 570.28 | 95.86 | 734.94 |
| PVC-U | pair | 3.80 | 60.80 | 525.26 | 87.91 | 673.97 |
| galvanised steel up-and-over garage doors | | | | | | |
| 2135 × 1980mm | nr | 6.00 | 96.00 | 316.69 | 61.90 | 474.59 |
| 3965 × 2135mm | nr | 7.50 | 120.00 | 1107.63 | 184.14 | 1411.77 |

| | Unit | Labour hours | Labour cost £ | Materials £ | O & P £ | Total £ |
|---|---|---|---|---|---|---|
| **Door repairs** | | | | | | |
| Take off and replace defective ironmongery to softwood doors | | | | | | |
| bolts | | | | | | |
| barrel | nr | 0.20 | 3.20 | 6.48 | 1.45 | 11.13 |
| flush | nr | 0.20 | 3.20 | 8.64 | 1.78 | 13.62 |
| tower | nr | 0.30 | 4.80 | 8.25 | 1.96 | 15.01 |
| butts | | | | | | |
| light | pair | 0.20 | 3.20 | 4.13 | 1.10 | 8.43 |
| medium | pair | 0.25 | 4.00 | 4.20 | 1.23 | 9.43 |
| heavy | pair | 0.30 | 4.80 | 5.28 | 1.51 | 11.59 |
| locks | | | | | | |
| cupboard | nr | 0.30 | 4.80 | 8.13 | 1.94 | 14.87 |
| mortice dead lock | nr | 0.85 | 13.60 | 14.04 | 4.15 | 31.79 |
| rim lock | nr | 0.45 | 7.20 | 6.35 | 2.03 | 15.58 |
| cylinder | nr | 1.10 | 17.60 | 27.00 | 6.69 | 51.29 |
| Take off and replace defective ironmongery to hardwood doors | | | | | | |
| bolts | | | | | | |
| barrel | nr | 0.30 | 4.80 | 6.48 | 1.69 | 12.97 |
| flush | nr | 0.30 | 4.80 | 8.64 | 2.02 | 15.46 |
| indicating | nr | 0.30 | 4.80 | 9.05 | 2.08 | 15.93 |
| tower | nr | 0.40 | 6.40 | 8.25 | 2.20 | 16.85 |
| butts | | | | | | |
| light | pair | 0.40 | 6.40 | 4.13 | 1.58 | 12.11 |
| medium | pair | 0.35 | 5.60 | 4.20 | 1.47 | 11.27 |
| heavy | pair | 0.40 | 6.40 | 5.28 | 1.75 | 13.43 |
| locks | | | | | | |
| cupboard | nr | 0.30 | 4.80 | 8.13 | 1.94 | 14.87 |
| mortice dead lock | nr | 0.85 | 13.60 | 14.04 | 4.15 | 31.79 |
| rim lock | nr | 0.45 | 7.20 | 6.35 | 2.03 | 15.58 |
| cylinder | nr | 1.10 | 17.60 | 27.00 | 6.69 | 51.29 |

| | Unit | Labour hours | Labour cost £ | Materials £ | O & P £ | Total £ |
|---|---|---|---|---|---|---|
| **Partitions, walls and ceilings** | | | | | | |
| Pull down existing damaged partitions and walls and rebuild stud partition plasterboard both sides | m2 | 1.50 | 24.00 | 10.18 | 5.13 | 39.31 |
| brickwork 112mm thick plastered both sides | m2 | 4.00 | 64.00 | 18.39 | 12.36 | 94.75 |
| blockwork 75mm thick plastered both sides | m2 | 3.10 | 49.60 | 11.21 | 9.12 | 69.93 |
| blockwork 100mm thick plastered both sides | m2 | 3.30 | 52.80 | 22.68 | 11.32 | 86.80 |
| Hack off scorched plaster to walls and renew | m2 | 1.50 | 24.00 | 2.11 | 3.92 | 30.03 |
| Pull down plasterboard and skim ceilings and renew | m2 | 1.56 | 24.96 | 3.81 | 4.32 | 33.09 |
| Hack off damaged wall tiles and renew | m2 | 1.45 | 23.20 | 25.92 | 7.37 | 56.49 |
| Take off damaged skirting and replace | | | | | | |
| 19 × 75mm softwood | m | 0.27 | 4.32 | 2.32 | 1.00 | 7.64 |
| 19 × 100mm softwood | m | 0.30 | 4.80 | 2.70 | 1.13 | 8.63 |
| 19 × 100mm hardwood | m | 0.32 | 5.12 | 6.12 | 1.69 | 12.93 |
| 25 × 150mm hardwood | m | 0.38 | 6.08 | 11.50 | 2.64 | 20.22 |
| Take up damaged flooring and renew | | | | | | |
| 19mm tongued and grooved softwood flooring | m2 | 1.14 | 18.24 | 11.69 | 4.49 | 34.42 |
| 25mm tongued and grooved softwood flooring | m2 | 1.15 | 18.40 | 13.76 | 4.82 | 36.98 |
| 12mm tongued and grooved chipboard flooring | m2 | 0.80 | 12.80 | 6.07 | 2.83 | 21.70 |
| 18mm tongued and grooved chipboard flooring | m2 | 1.00 | 16.00 | 6.52 | 3.38 | 25.90 |
| 2mm vinyl sheeting | m2 | 0.50 | 8.00 | 11.02 | 2.85 | 21.87 |

| | Unit | Labour hours | Labour cost £ | Materials £ | O & P £ | Total £ |
|---|---|---|---|---|---|---|
| 2.5mm vinyl sheeting | m2 | 0.55 | 8.80 | 14.79 | 3.54 | 27.13 |
| 3mm vinyl sheeting | m2 | 0.60 | 9.60 | 15.06 | 3.70 | 28.36 |
| 2mm vinyl tiling | m2 | 0.40 | 6.40 | 9.69 | 2.41 | 18.50 |
| 3mm laminated floor | m2 | 0.90 | 14.40 | 11.79 | 3.93 | 30.12 |
| 12.5mm quarry tiling | m2 | 1.15 | 18.40 | 35.56 | 8.09 | 62.05 |

Cut out and replace floor joists and roof members

| | Unit | Labour hours | Labour cost £ | Materials £ | O & P £ | Total £ |
|---|---|---|---|---|---|---|
| 38 × 100mm softwood | m | 0.22 | 3.52 | 1.51 | 0.75 | 5.78 |
| 50 × 75mm softwood | m | 0.24 | 3.84 | 1.98 | 0.87 | 6.69 |
| 50 × 100mm softwood | m | 0.28 | 4.48 | 2.38 | 1.03 | 7.89 |
| 50 × 125mm softwood | m | 0.30 | 4.80 | 2.60 | 1.11 | 8.51 |
| 50 × 150mm softwood | m | 0.32 | 5.12 | 2.82 | 1.19 | 9.13 |
| 75 × 125mm softwood | m | 0.34 | 5.44 | 3.32 | 1.31 | 10.07 |
| 75 × 200mm softwood | m | 0.36 | 5.76 | 4.06 | 1.47 | 11.29 |

Take down damaged softwood staircase 2600mm rise and renew

| | Unit | Labour hours | Labour cost £ | Materials £ | O & P £ | Total £ |
|---|---|---|---|---|---|---|
| straight flight 900mm wide | nr | 12.20 | 195.20 | 681.74 | 131.54 | 1008.48 |
| straight flight 900mm wide with balustrade | nr | 13.30 | 212.80 | 783.40 | 149.43 | 1145.63 |
| two flights 900mm wide with landing and balustrade | nr | 14.50 | 232.00 | 874.66 | 166.00 | 1272.66 |

Take down damaged hardwood staircase 2600mm rise and renew

| | Unit | Labour hours | Labour cost £ | Materials £ | O & P £ | Total £ |
|---|---|---|---|---|---|---|
| straight flight 900mm wide | nr | 12.40 | 198.40 | 1155.60 | 203.10 | 1557.10 |
| straight flight 900mm wide with balustrade | nr | 13.90 | 222.40 | 1261.44 | 222.58 | 1706.42 |
| two flights 900mm wide with landing and balustrade | nr | 15.00 | 240.00 | 1391.04 | 244.66 | 1875.70 |

| | Unit | Labour hours | Labour cost £ | Materials £ | O & P £ | Total £ |
|---|---|---|---|---|---|---|
| **Furniture and fittings** | | | | | | |
| Take down damaged fittings and replace (material prices not included due to the wide variation in the quality and costs of the fittings) | | | | | | |
| wall units 2000mm high | | | | | | |
| 300 × 600mm | nr | 1.00 | 16.00 | 0.00 | 2.40 | 18.40 |
| 300 × 1000mm | nr | 1.10 | 17.60 | 0.00 | 2.64 | 20.24 |
| 300 × 1200mm | nr | 1.20 | 19.20 | 0.00 | 2.88 | 22.08 |
| base units 750mm high | | | | | | |
| 600 × 900mm | nr | 1.00 | 16.00 | 0.00 | 2.40 | 18.40 |
| 600 × 1000mm | nr | 1.00 | 16.00 | 0.00 | 2.40 | 18.40 |
| 600 × 1200mm | nr | 1.10 | 17.60 | 0.00 | 2.64 | 20.24 |
| 900 × 900mm | nr | 1.10 | 17.60 | 0.00 | 2.64 | 20.24 |
| 900 × 1000mm | nr | 1.20 | 19.20 | 0.00 | 2.88 | 22.08 |
| 900 × 1200mm | nr | 1.20 | 19.20 | 0.00 | 2.88 | 22.08 |
| sink units 750mm high | | | | | | |
| 600 × 900mm | nr | 1.40 | 22.40 | 0.00 | 3.36 | 25.76 |
| 600 × 1000mm | nr | 1.40 | 22.40 | 0.00 | 3.36 | 25.76 |
| 600 × 1200mm | nr | 1.40 | 22.40 | 0.00 | 3.36 | 25.76 |
| worktops | | | | | | |
| 600 × 900mm | nr | 0.80 | 12.80 | 0.00 | 1.92 | 14.72 |
| 600 × 1000mm | nr | 0.90 | 14.40 | 0.00 | 2.16 | 16.56 |
| 600 × 1200mm | nr | 1.00 | 16.00 | 0.00 | 2.40 | 18.40 |
| Remove smoke damaged furniture and place in skip | | | | | | |
| easy chair | nr | 0.15 | 2.40 | 0.00 | 0.36 | 2.76 |
| settee | nr | 0.18 | 2.88 | 0.00 | 0.43 | 3.31 |
| single bed/mattress | nr | 0.20 | 3.20 | 0.00 | 0.48 | 3.68 |
| double bed/mattress | nr | 0.22 | 3.52 | 0.00 | 0.53 | 4.05 |
| set of six dining chairs | nr | 0.12 | 1.92 | 0.00 | 0.29 | 2.21 |
| dining table | nr | 0.18 | 2.88 | 0.00 | 0.43 | 3.31 |
| sideboard | nr | 0.20 | 3.20 | 0.00 | 0.48 | 3.68 |

| | Unit | Labour hours | Labour cost £ | Materials £ | O & P £ | Total £ |
|---|---|---|---|---|---|---|
| **Remove smoke damaged furniture (cont'd)** | | | | | | |
| wardrobe | nr | 0.15 | 2.40 | 0.00 | 0.36 | 2.76 |
| chest of drawers | nr | 0.20 | 3.20 | 0.00 | 0.48 | 3.68 |
| carpet | nr | 0.25 | 4.00 | 0.00 | 0.60 | 4.60 |
| TV/hi-fi equipment | nr | 0.25 | 4.00 | 0.00 | 0.60 | 4.60 |
| **Plumbing and heating work** | | | | | | |
| **Take out damaged sanitary fittings and associated pipework and replace** | | | | | | |
| lavatory basin | nr | 3.30 | 52.80 | 142.56 | 29.30 | 224.66 |
| low level WC | nr | 3.40 | 54.40 | 361.80 | 62.43 | 478.63 |
| shower cubicle | nr | 4.20 | 67.20 | 986.24 | 158.02 | 1211.46 |
| sink/single drainer | nr | 2.30 | 36.80 | 199.80 | 35.49 | 272.09 |
| sink/double drainer | nr | 2.90 | 46.40 | 251.64 | 44.71 | 342.75 |
| bath, acrylic | nr | 4.20 | 67.20 | 307.80 | 56.25 | 431.25 |
| bidet | nr | 3.50 | 56.00 | 424.44 | 72.07 | 552.51 |
| **Take out damaged cisterns and cylinders and associated pipework and replace** | | | | | | |
| polyethylene cold water cisterns | | | | | | |
| 68 litres | nr | 2.00 | 32.00 | 97.20 | 19.38 | 148.58 |
| 86 litres | nr | 2.00 | 32.00 | 109.08 | 21.16 | 162.24 |
| 191 litres | nr | 2.50 | 40.00 | 145.80 | 27.87 | 213.67 |
| 327 litres | nr | 3.80 | 60.80 | 179.28 | 36.01 | 276.09 |
| copper cylinders, indirect pattern | | | | | | |
| 114 litres | nr | 3.50 | 56.00 | 139.32 | 29.30 | 224.62 |
| 117 litres | nr | 3.50 | 56.00 | 147.96 | 30.59 | 234.55 |
| 140 litres | nr | 4.00 | 64.00 | 162.00 | 33.90 | 259.90 |
| 162 litres | nr | 4.10 | 65.60 | 198.72 | 39.65 | 303.97 |

| | Unit | Labour hours | Labour cost £ | Materials £ | O & P £ | Total £ |
|---|---|---|---|---|---|---|
| copper cylinders, direct pattern | | | | | | |
| 116 litres | nr | 3.80 | 60.80 | 136.08 | 29.53 | 226.41 |
| 120 litres | nr | 3.80 | 60.80 | 153.36 | 32.12 | 246.28 |
| 144 litres | nr | 4.20 | 67.20 | 163.08 | 34.54 | 264.82 |
| 166 litres | nr | 4.30 | 68.80 | 183.60 | 37.86 | 290.26 |
| **Disconnect damaged central heating boiler and associated pipe work and replace** | | | | | | |
| floor-mounted gas boiler | | | | | | |
| 30,000 Btu | nr | 9.00 | 144.00 | 648.00 | 118.80 | 910.80 |
| 40,000 Btu | nr | 9.00 | 144.00 | 679.32 | 123.50 | 946.82 |
| 50,000 Btu | nr | 9.20 | 147.20 | 762.48 | 136.45 | 1046.13 |
| 60,000 Btu | nr | 9.20 | 147.20 | 858.60 | 150.87 | 1156.67 |
| wall-mounted gas boiler | | | | | | |
| 30,000 Btu | nr | 9.00 | 144.00 | 719.28 | 129.49 | 992.77 |
| 40,000 Btu | nr | 9.00 | 144.00 | 775.44 | 137.92 | 1057.36 |
| 50,000 Btu | nr | 9.20 | 147.20 | 843.48 | 148.60 | 1139.28 |
| 60,000 Btu | nr | 9.20 | 147.20 | 908.28 | 158.32 | 1213.80 |
| floor-mounted oil-fired boiler | | | | | | |
| 52,000 Btu | nr | 9.20 | 147.20 | 1503.86 | 247.66 | 1898.72 |
| 70,000 Btu | nr | 9.40 | 150.40 | 1629.27 | 266.95 | 2046.62 |
| **Disconnect damaged radiator and associated pipework and and valves and replace** | | | | | | |
| single panel, 450mm high | | | | | | |
| length, 1000mm | nr | 1.65 | 26.40 | 77.76 | 15.62 | 119.78 |
| length, 1600mm | nr | 1.85 | 29.60 | 109.08 | 20.80 | 159.48 |
| length, 2000mm | nr | 2.10 | 33.60 | 129.60 | 24.48 | 187.68 |
| single panel, 600mm high | | | | | | |
| length, 1000mm | nr | 1.85 | 29.60 | 96.12 | 18.86 | 144.58 |
| length, 1600mm | nr | 2.05 | 32.80 | 135.00 | 25.17 | 192.97 |
| length, 2000mm | nr | 2.30 | 36.80 | 162.23 | 29.85 | 228.88 |

| | Unit | Labour hours | Labour cost £ | Materials £ | O & P £ | Total £ |
|---|---|---|---|---|---|---|
| double panel, 450mm high | | | | | | |
| length, 1000mm | nr | 1.85 | 29.60 | 111.24 | 21.13 | 161.97 |
| length, 1400mm | nr | 1.35 | 21.60 | 151.20 | 25.92 | 198.72 |
| length, 2000mm | nr | 1.45 | 23.20 | 192.24 | 32.32 | 247.76 |
| double panel, 600mm high | | | | | | |
| length, 1000mm | nr | 2.05 | 32.80 | 108.81 | 21.24 | 162.85 |
| length, 1400mm | nr | 2.25 | 36.00 | 153.36 | 28.40 | 217.76 |
| length, 2000mm | nr | 2.45 | 39.20 | 262.44 | 45.25 | 346.89 |

## Electrics

Disconnect power supply, remove damaged electric fittings and replace

| | Unit | Labour hours | Labour cost £ | Materials £ | O & P £ | Total £ |
|---|---|---|---|---|---|---|
| single power point | nr | 0.60 | 9.60 | 7.64 | 2.59 | 19.83 |
| double power point | nr | 0.60 | 9.60 | 8.39 | 2.70 | 20.69 |
| light switch | nr | 0.60 | 9.60 | 6.36 | 2.39 | 18.35 |
| light point | nr | 0.60 | 9.60 | 7.11 | 2.51 | 19.22 |
| wall light | nr | 0.90 | 14.40 | 25.04 | 5.92 | 45.36 |
| cooker control unit | nr | 1.50 | 24.00 | 20.51 | 6.68 | 51.19 |

## Painting and decorating

Burn off damaged woodwork and leave ready to receive new paint-work

| | Unit | Labour hours | Labour cost £ | Materials £ | O & P £ | Total £ |
|---|---|---|---|---|---|---|
| general surfaces | m2 | 1.10 | 17.60 | 0.00 | 2.64 | 20.24 |
| general surfaces up to 300mm wide | m | 0.30 | 4.80 | 0.00 | 0.72 | 5.52 |

Burn off damaged metalwork and leave ready to receive new paint-work

| | Unit | Labour hours | Labour cost £ | Materials £ | O & P £ | Total £ |
|---|---|---|---|---|---|---|
| general surfaces | m2 | 1.10 | 17.60 | 0.00 | 2.64 | 20.24 |
| general surfaces up to 300mm wide | m | 0.30 | 4.80 | 0.00 | 0.72 | 5.52 |

| | Unit | Labour hours | Labour cost £ | Materials £ | O & P £ | Total £ |
|---|---|---|---|---|---|---|
| Wash down existing plastered surfaces, stop cracks and rub down and leave ready to receive new paintwork | | | | | | |
| walls | m2 | 0.14 | 2.24 | 0.00 | 0.34 | 2.58 |
| ceilings | m2 | 0.05 | 0.80 | 0.00 | 0.12 | 0.92 |
| Apply one mist coat and two coats emulsion paint to plastered surfaces | | | | | | |
| walls | m2 | 0.30 | 4.80 | 1.16 | 0.89 | 6.85 |
| ceilings | m2 | 0.32 | 5.12 | 1.16 | 0.94 | 7.22 |
| Apply one undercoat and one coat gloss paint to plastered surfaces | | | | | | |
| walls | m2 | 0.30 | 4.80 | 1.52 | 0.95 | 7.27 |
| ceilings | m2 | 0.36 | 5.76 | 1.52 | 1.09 | 8.37 |
| Prepare, size, apply adhesive, supply and hang paper to plastered walls | | | | | | |
| lining paper | | | | | | |
| £1.50 per roll | m2 | 0.30 | 4.80 | 0.36 | 0.77 | 5.93 |
| £2.00 per roll | m2 | 0.30 | 4.80 | 0.48 | 0.79 | 6.07 |
| £2.50 per roll | m2 | 0.30 | 4.80 | 0.59 | 0.81 | 6.20 |
| washable paper | | | | | | |
| £2.50 per roll | m2 | 0.30 | 4.80 | 1.18 | 0.90 | 6.88 |
| £4.00 per roll | m2 | 0.30 | 4.80 | 1.41 | 0.93 | 7.14 |
| £5.00 per roll | m2 | 0.30 | 4.80 | 1.64 | 0.97 | 7.41 |
| vinyl paper | | | | | | |
| £4.00 per roll | m2 | 0.30 | 4.80 | 0.94 | 0.86 | 6.60 |
| £5.00 per roll | m2 | 0.30 | 4.80 | 1.18 | 0.90 | 6.88 |
| £6.00 per roll | m2 | 0.30 | 4.80 | 1.41 | 0.93 | 7.14 |

| | Unit | Labour hours | Labour cost £ | Materials £ | O & P £ | Total £ |
|---|---|---|---|---|---|---|
| **FLOOD DAMAGE** | | | | | | |
| **Pumping** | | | | | | |
| Instal pump and hoses in position | Item | 1.00 | 16.00 | 0.00 | 2.40 | 18.40 |
| Hire diaphragm pump and hoses | | | | | | |
| 50mm | day | 1.00 | 16.00 | 45.00 | 9.15 | 70.15 |
| 75mm | day | 1.00 | 16.00 | 55.00 | 10.65 | 81.65 |
| Hire portable dryer, 350W | week | 1.00 | 16.00 | 70.00 | 12.90 | 98.90 |
| Clean up and remove debris from basement and ground floor | m2 | 0.20 | 3.20 | 0.00 | 0.48 | 3.68 |
| Hack off defective wall plaster and replace | m2 | 0.20 | 3.20 | 0.00 | 0.48 | 3.68 |
| Pull down plasterboard and skim ceilings and renew | m2 | 1.56 | 24.96 | 3.82 | 4.32 | 33.10 |
| Hack off damaged wall tiles and renew | m2 | 1.45 | 23.20 | 26.49 | 7.45 | 57.14 |
| Take off damaged skirting and replace | | | | | | |
| 19 × 100mm softwood | m | 0.30 | 4.80 | 2.71 | 1.13 | 8.64 |
| 25 × 150mm hardwood | m | 0.38 | 6.08 | 11.55 | 2.64 | 20.27 |
| Take up damaged flooring and renew | | | | | | |
| 19mm tongued and grooved | | | | | | |
| softwood flooring | m2 | 1.14 | 18.24 | 11.74 | 4.50 | 34.48 |
| chipboard flooring | m2 | 0.80 | 12.80 | 6.09 | 2.83 | 21.72 |

| | Unit | Labour hours | Labour cost £ | Materials £ | O & P £ | Total £ |
|---|---|---|---|---|---|---|
| 18mm tongued and grooved chipboard flooring | m2 | 1.00 | 16.00 | 6.09 | 3.31 | 25.40 |

**Electrics**

| | | | | | | |
|---|---|---|---|---|---|---|
| Disconnect power supply, remove damaged electric fittings and replace | | | | | | |
| single power point | nr | 0.60 | 9.60 | 7.66 | 2.59 | 19.85 |
| double power point | nr | 0.60 | 9.60 | 8.42 | 2.70 | 20.72 |
| light switch | nr | 0.60 | 9.60 | 6.38 | 2.40 | 18.38 |
| cooker control unit | nr | 1.50 | 24.00 | 20.58 | 6.69 | 51.27 |

**Painting and decorating**

| | | | | | | |
|---|---|---|---|---|---|---|
| Burn off damaged woodwork and leave ready to receive new paintwork | | | | | | |
| general surfaces | m2 | 1.10 | 17.60 | 0.00 | 2.64 | 20.24 |
| general surfaces up to 300mm wide | m | 0.30 | 4.80 | 0.00 | 0.72 | 5.52 |
| Burn off damaged metalwork and leave ready to receive new paintwork | | | | | | |
| general surfaces | m2 | 1.10 | 17.60 | 0.00 | 2.64 | 20.24 |
| general surfaces up to 300mm wide | m | 0.30 | 4.80 | 0.00 | 0.72 | 5.52 |
| Wash down existing plastered surfaces, stop cracks and rub down and leave ready to receive new paintwork | | | | | | |
| walls | m2 | 0.05 | 0.80 | 0.00 | 0.12 | 0.92 |
| ceilings | m2 | 0.07 | 1.12 | 0.00 | 0.17 | 1.29 |
| Apply one undercoat and one coat gloss paint to plastered surfaces | | | | | | |
| walls | m2 | 0.30 | 4.80 | 1.52 | 0.95 | 7.27 |
| ceilings | m2 | 0.36 | 5.76 | 1.52 | 1.09 | 8.37 |

|  | Unit | Labour hours | Labour cost £ | Materials £ | O & P £ | Total £ |
|---|---|---|---|---|---|---|
| **GALE DAMAGE** | | | | | | |
| **Hoardings and screens** | | | | | | |
| Temporary screens and hoardings 2m high consisting of 22mm thick exterior quality plywood fixed to 100 × 50mm posts and rails | m | 2.60 | 41.60 | 36.26 | 11.68 | 89.54 |
| Tarpaulin sheeting size 5 × 4m fixed in position | week | 0.00 | 0.00 | 17.00 | 2.55 | 19.55 |
| Plywood sheeting blocking up window or door opening | m2 | 1.00 | 16.00 | 18.22 | 5.13 | 39.35 |
| **Shoring** | | | | | | |
| Timber dead shores consisting of 200 × 200mm shores, 200 × 50mm plates and 200 × 50mm braces at centres of | | | | | | |
| 2m | m2 | 3.80 | 60.80 | 86.25 | 22.06 | 169.11 |
| 3m | m2 | 3.00 | 48.00 | 17.22 | 9.78 | 75.00 |
| 4m | m2 | 2.20 | 35.20 | 12.50 | 7.16 | 54.86 |
| Timber raking and flying shores consisting of 200 × 200mm shores, 250 × 50mm plates and 200 × 50mm braces to gable end of two-storey house | Item | 70.00 | 1120.00 | 560.99 | 252.15 | 1933.14 |
| **Roof repairs** | | | | | | |
| Take up roof coverings from pitched roof | | | | | | |
| tiles | m2 | 0.80 | 12.80 | 0.00 | 1.92 | 14.72 |
| slates | m2 | 0.80 | 12.80 | 0.00 | 1.92 | 14.72 |
| timber boarding | m2 | 1.00 | 16.00 | 0.00 | 2.40 | 18.40 |
| flat sheeting | m2 | 0.30 | 4.80 | 0.00 | 0.72 | 5.52 |

| | Unit | Labour hours | Labour cost £ | Materials £ | O & P £ | Total £ |
|---|---|---|---|---|---|---|
| **Take up roof coverings from flat roof** | | | | | | |
| bituminous felt | m2 | 0.25 | 4.00 | 0.00 | 0.60 | 4.60 |
| metal sheeting | m2 | 0.30 | 4.80 | 0.00 | 0.72 | 5.52 |
| woodwool slabs | m2 | 0.50 | 8.00 | 0.00 | 1.20 | 9.20 |
| firrings | m2 | 0.20 | 3.20 | 0.00 | 0.48 | 3.68 |
| **Take up roof coverings from pitched roof, carefully lay aside for reuse** | | | | | | |
| tiles | m2 | 1.10 | 17.60 | 0.00 | 2.64 | 20.24 |
| slates | m2 | 1.10 | 17.60 | 0.00 | 2.64 | 20.24 |
| metal sheeting | m2 | 0.50 | 8.00 | 0.00 | 1.20 | 9.20 |
| flat sheeting | m2 | 0.60 | 9.60 | 0.00 | 1.44 | 11.04 |
| corrugated sheeting | m2 | 0.60 | 9.60 | 0.00 | 1.44 | 11.04 |
| **Take up roof coverings from flat roof, carefully lay aside for reuse** | | | | | | |
| metal sheeting | m2 | 0.60 | 9.60 | 0.00 | 1.44 | 11.04 |
| woodwool slabs | m2 | 0.80 | 12.80 | 0.00 | 1.92 | 14.72 |
| **Inspect roof battens, refix loose and replace with new, size 38 × 25mm** | | | | | | |
| **25% of area** | | | | | | |
| 250mm centres | m2 | 0.14 | 2.24 | 0.50 | 0.41 | 3.15 |
| 450mm centres | m2 | 0.12 | 1.92 | 0.44 | 0.35 | 2.71 |
| 600mm centres | m2 | 0.10 | 1.60 | 0.37 | 0.30 | 2.27 |
| **75% of area** | | | | | | |
| 250mm centres | m2 | 0.36 | 5.76 | 1.27 | 1.05 | 8.08 |
| 450mm centres | m2 | 0.26 | 4.16 | 1.08 | 0.79 | 6.03 |
| 600mm centres | m2 | 0.22 | 3.52 | 0.88 | 0.66 | 5.06 |
| **100% of area** | | | | | | |
| 250mm centres | m2 | 0.44 | 7.04 | 0.55 | 1.14 | 8.73 |
| 450mm centres | m2 | 0.32 | 5.12 | 1.39 | 0.98 | 7.49 |
| 600mm centres | m2 | 0.38 | 6.08 | 1.24 | 1.10 | 8.42 |

| | Unit | Labour hours | Labour cost £ | Materials £ | O & P £ | Total £ |
|---|---|---|---|---|---|---|
| Remove slate and refix | nr | 1.00 | 16.00 | 1.13 | 2.57 | 19.70 |
| **Remove single broken slate, renew with new Welsh blue slate** | | | | | | |
| 405 × 255mm | nr | 1.20 | 19.20 | 1.65 | 3.13 | 23.98 |
| 510 × 255mm | nr | 1.20 | 19.20 | 3.17 | 3.36 | 25.73 |
| 610 × 305mm | nr | 1.20 | 19.20 | 6.50 | 3.86 | 29.56 |
| **Remove slates in area approximately 1m2 and replace with Welsh slates previously laid aside** | nr | 1.60 | 25.60 | 0.00 | 3.84 | 29.44 |
| **Remove slates in area approximately 1m2 and replace with Welsh blue slates previously laid aside** | | | | | | |
| 405 × 255mm | nr | 1.80 | 28.80 | 0.00 | 4.32 | 33.12 |
| 510 × 255mm | nr | 1.70 | 27.20 | 0.00 | 4.08 | 31.28 |
| 610 × 305mm | nr | 1.60 | 25.60 | 0.00 | 3.84 | 29.44 |
| **Remove double course at eaves and fix new Welsh blue slates** | | | | | | |
| 405 × 255mm | m | 0.70 | 11.20 | 56.05 | 10.09 | 77.34 |
| 510 × 255mm | m | 0.70 | 11.20 | 58.41 | 10.44 | 80.05 |
| 610 × 305mm | m | 0.70 | 11.20 | 64.20 | 11.31 | 86.71 |
| **Remove single slipped tile and refix** | nr | 0.30 | 4.80 | 0.00 | 0.72 | 5.52 |
| **Remove single broken tile and renew** | | | | | | |
| Marley plain tile | nr | 0.30 | 4.80 | 0.83 | 0.84 | 6.47 |
| Marley Ludlow Plus | nr | 0.30 | 4.80 | 1.06 | 0.88 | 6.74 |
| Marley Modern tile | nr | 0.30 | 4.80 | 1.71 | 0.98 | 7.49 |
| Redland Norfolk tile | nr | 0.30 | 4.80 | 1.43 | 0.93 | 7.16 |
| **Remove tiles in area approximately 1m2 and replace with tiles previously laid aside** | | | | | | |
| Marley Plain tile | nr | 1.80 | 28.80 | 35.45 | 9.64 | 73.89 |
| Marley Ludlow Plus | nr | 1.20 | 19.20 | 12.34 | 4.73 | 36.27 |
| Marley Modern tile | nr | 1.10 | 17.60 | 19.50 | 5.57 | 42.67 |

| | Unit | Labour hours | Labour cost £ | Materials £ | O & P £ | Total £ |
|---|---|---|---|---|---|---|
| Take off defective ridge capping and refix including pointing in mortar | m | 1.10 | 17.60 | 1.67 | 2.89 | 22.16 |
| **Chimney stack** | | | | | | |
| Erect and take down chimney scaffold | item | 4.00 | 64.00 | 0.00 | 9.60 | 73.60 |
| Hire chimney scaffold and platform | week | 0.00 | 0.00 | 98.00 | 14.70 | 112.70 |
| Take down existing chimney stack to below roof level and remove debris | | | | | | |
|   single stack | m | 2.50 | 40.00 | 1.52 | 6.23 | 47.75 |
|   double stack | m | 3.50 | 56.00 | 1.52 | 8.63 | 66.15 |
| Chimney stack in facing brick £450.00 per thousand in gauged mortar | | | | | | |
|   single stack | m | 3.80 | 60.80 | 42.27 | 15.46 | 118.53 |
|   double stack | m | 5.20 | 83.20 | 102.48 | 27.85 | 213.53 |
| Terra cotta chimney pot bedded and flaunched in gauged mortar | | | | | | |
|   185mm diameter × 300mm high | nr | 1.25 | 20.00 | 30.55 | 7.58 | 58.13 |
|   185mm diameter × 600mm high | nr | 1.80 | 28.80 | 52.50 | 12.20 | 93.50 |
| **External work** | | | | | | |
| Remove blown-down trees, trunk girth 1m above ground level | | | | | | |
|   600 to 1500mm | nr | 22.00 | 352.00 | 0.00 | 52.80 | 404.80 |
|   1500 to 3000mm | nr | 38.00 | 608.00 | 0.00 | 91.20 | 699.20 |

| | Unit | Labour hours | Labour cost £ | Materials £ | O & P £ | Total £ |
|---|---|---|---|---|---|---|

**THEFT DAMAGE**

**Window and door repairs**

| | Unit | Labour hours | Labour cost £ | Materials £ | O & P £ | Total £ |
|---|---|---|---|---|---|---|
| Hack out glass and remove | m2 | 0.45 | 7.20 | 0.00 | 1.08 | 8.28 |
| Clean rebates, remove sprigs or clips and prepare for reglazing | m | 0.20 | 3.20 | 0.00 | 0.48 | 3.68 |

Reglaze existing softwood windows in clear float glass with putty and sprigs

under 0.15m2, thickness

| | Unit | Labour hours | Labour cost £ | Materials £ | O & P £ | Total £ |
|---|---|---|---|---|---|---|
| 4mm | m2 | 0.90 | 14.40 | 38.02 | 7.86 | 60.28 |
| 5mm | m2 | 0.90 | 14.40 | 54.21 | 10.29 | 78.90 |
| 6mm | m2 | 1.00 | 16.00 | 56.08 | 10.81 | 82.89 |
| 10mm | m2 | 1.05 | 16.80 | 108.91 | 18.86 | 144.57 |

over 0.15m2, thickness

| | Unit | Labour hours | Labour cost £ | Materials £ | O & P £ | Total £ |
|---|---|---|---|---|---|---|
| 4mm | m2 | 0.60 | 9.60 | 38.02 | 7.14 | 54.76 |
| 5mm | m2 | 0.60 | 9.60 | 54.21 | 9.57 | 73.38 |
| 6mm | m2 | 0.65 | 10.40 | 56.08 | 9.97 | 76.45 |
| 10mm | m2 | 0.70 | 11.20 | 108.91 | 18.02 | 138.13 |

Reglaze existing softwood windows in clear float glass with pinned beads

under 0.15m2, thickness

| | Unit | Labour hours | Labour cost £ | Materials £ | O & P £ | Total £ |
|---|---|---|---|---|---|---|
| 4mm | m2 | 1.10 | 17.60 | 38.40 | 8.40 | 64.40 |
| 5mm | m2 | 1.10 | 17.60 | 57.75 | 11.30 | 86.65 |
| 6mm | m2 | 1.20 | 19.20 | 68.75 | 13.19 | 101.14 |
| 10mm | m2 | 1.25 | 20.00 | 109.98 | 19.50 | 149.48 |

over 0.15m2, thickness

| | Unit | Labour hours | Labour cost £ | Materials £ | O & P £ | Total £ |
|---|---|---|---|---|---|---|
| 4mm | m2 | 0.80 | 12.80 | 38.40 | 7.68 | 58.88 |
| 5mm | m2 | 0.80 | 12.80 | 57.75 | 10.58 | 81.13 |
| 6mm | m2 | 0.90 | 14.40 | 68.75 | 12.47 | 95.62 |
| 10mm | m2 | 0.95 | 15.20 | 109.98 | 18.78 | 143.96 |

| | Unit | Labour hours | Labour cost £ | Materials £ | O & P £ | Total £ |
|---|---|---|---|---|---|---|
| **Reglaze existing metal windows in clear float glass with clips and putty** | | | | | | |
| under 0.15m2, thickness | | | | | | |
| 4mm | m2 | 0.95 | 15.20 | 38.02 | 7.98 | 61.20 |
| 5mm | m2 | 0.95 | 15.20 | 54.21 | 10.41 | 79.82 |
| 6mm | m2 | 1.05 | 16.80 | 56.08 | 10.93 | 83.81 |
| 10mm | m2 | 1.10 | 17.60 | 108.91 | 18.98 | 145.49 |
| over 0.15m2, thickness | | | | | | |
| 4mm | m2 | 0.65 | 10.40 | 38.02 | 7.26 | 55.68 |
| 5mm | m2 | 0.65 | 10.40 | 54.21 | 9.69 | 74.30 |
| 6mm | m2 | 0.70 | 11.20 | 56.08 | 10.09 | 77.37 |
| 10mm | m2 | 0.75 | 12.00 | 108.91 | 18.14 | 139.05 |
| **Reglaze existing metal windows in clear float glass with screwed metal beads** | | | | | | |
| under 0.15m2, thickness | | | | | | |
| 4mm | m2 | 1.30 | 20.80 | 38.02 | 8.82 | 67.64 |
| 5mm | m2 | 1.30 | 20.80 | 54.21 | 11.25 | 86.26 |
| 6mm | m2 | 1.40 | 22.40 | 56.08 | 11.77 | 90.25 |
| 10mm | m2 | 1.45 | 23.20 | 108.91 | 19.82 | 151.93 |
| over 0.15m2, thickness | | | | | | |
| 4mm | m2 | 1.00 | 16.00 | 38.02 | 8.10 | 62.12 |
| 5mm | m2 | 1.00 | 16.00 | 54.21 | 10.53 | 80.74 |
| 6mm | m2 | 1.05 | 16.80 | 56.08 | 10.93 | 83.81 |
| 10mm | m2 | 1.10 | 17.60 | 108.91 | 18.98 | 145.49 |
| **Reglaze existing softwood windows in white patterned glass with putty and sprigs** | | | | | | |
| under 0.15m2, thickness | | | | | | |
| 4mm | m2 | 0.90 | 14.40 | 37.20 | 7.74 | 59.34 |
| 6mm | m2 | 1.00 | 16.00 | 56.16 | 10.82 | 82.98 |
| over 0.15m2, thickness | | | | | | |
| 4mm | m2 | 0.60 | 9.60 | 37.20 | 7.02 | 53.82 |
| 6mm | m2 | 0.65 | 10.40 | 56.16 | 9.98 | 76.54 |

| | Unit | Labour hours | Labour cost £ | Materials £ | O & P £ | Total £ |
|---|---|---|---|---|---|---|
| **Reglaze existing softwood windows in white patterned glass with pinned beads** | | | | | | |
| under 0.15m2, thickness | | | | | | |
| 4mm | m2 | 1.10 | 17.60 | 37.20 | 8.22 | 63.02 |
| 6mm | m2 | 1.20 | 19.20 | 56.16 | 11.30 | 86.66 |
| over 0.15m2, thickness | | | | | | |
| 4mm | m2 | 0.80 | 12.80 | 37.20 | 7.50 | 57.50 |
| 6mm | m2 | 0.85 | 13.60 | 56.16 | 10.46 | 80.22 |
| **Reglaze existing softwood windows in white patterned glass with screwed beads** | | | | | | |
| under 0.15m2, thickness | | | | | | |
| 4mm | m2 | 1.30 | 20.80 | 37.20 | 8.70 | 66.70 |
| 6mm | m2 | 1.40 | 22.40 | 56.16 | 11.78 | 90.34 |
| over 0.15m2, thickness | | | | | | |
| 4mm | m2 | 1.00 | 16.00 | 37.20 | 7.98 | 61.18 |
| 6mm | m2 | 1.50 | 24.00 | 56.16 | 12.02 | 92.18 |
| **Reglaze existing metal windows in white patterned glass with putty** | | | | | | |
| under 0.15m2, thickness | | | | | | |
| 4mm | m2 | 0.90 | 14.40 | 37.20 | 7.74 | 59.34 |
| 6mm | m2 | 1.00 | 16.00 | 56.16 | 10.82 | 82.98 |
| over 0.15m2, thickness | | | | | | |
| 4mm | m2 | 0.60 | 9.60 | 37.20 | 7.02 | 53.82 |
| 6mm | m2 | 0.65 | 10.40 | 56.16 | 9.98 | 76.54 |
| **Reglaze existing metal windows in white patterned glass with metal clips and putty** | | | | | | |
| under 0.15m2, thickness | | | | | | |
| 4mm | m2 | 0.95 | 15.20 | 37.20 | 7.86 | 60.26 |
| 6mm | m2 | 1.05 | 16.80 | 56.16 | 10.94 | 83.90 |

| | Unit | Labour hours | Labour cost £ | Materials £ | O & P £ | Total £ |
|---|---|---|---|---|---|---|
| over 0.15m2, thickness | | | | | | |
| 4mm | m2 | 0.65 | 10.40 | 37.20 | 7.14 | 54.74 |
| 6mm | m2 | 0.70 | 11.20 | 56.16 | 10.10 | 77.46 |
| Reglaze existing metal windows in white patterned glass with screwed beads | | | | | | |
| under 0.15m2, thickness | | | | | | |
| 4mm | m2 | 1.30 | 20.80 | 37.20 | 8.70 | 66.70 |
| 6mm | m2 | 1.40 | 22.40 | 56.16 | 11.78 | 90.34 |
| over 0.15m2, thickness | | | | | | |
| 4mm | m2 | 1.00 | 16.00 | 37.20 | 7.98 | 61.18 |
| 6mm | m2 | 1.05 | 16.80 | 56.16 | 10.94 | 83.90 |
| Take off and replace damaged ironmongery to softwood windows | | | | | | |
| casement fastener | nr | 0.20 | 3.20 | 4.59 | 1.17 | 8.96 |
| casement stay | nr | 0.20 | 3.20 | 6.98 | 1.53 | 11.71 |
| hinges | pair | 0.25 | 4.00 | 4.17 | 1.23 | 9.40 |
| lock | nr | 0.60 | 9.60 | 14.16 | 3.56 | 27.32 |
| Take off and replace damaged ironmongery to hardwood windows | | | | | | |
| casement fastener | nr | 0.30 | 4.80 | 4.59 | 1.41 | 10.80 |
| casement stay | nr | 0.30 | 4.80 | 6.98 | 1.77 | 13.55 |
| hinges | pair | 0.35 | 5.60 | 4.17 | 1.47 | 11.24 |
| lock | nr | 0.75 | 12.00 | 14.16 | 3.92 | 30.08 |
| Take off and replace defective ironmongery to PVC-U windows | | | | | | |
| casement fastener | nr | 0.20 | 3.20 | 4.59 | 1.17 | 8.96 |
| casement stay | nr | 0.20 | 3.20 | 6.98 | 1.53 | 11.71 |
| hinges | pair | 0.25 | 4.00 | 4.17 | 1.23 | 9.40 |
| lock | nr | 0.60 | 9.60 | 14.16 | 3.56 | 27.32 |
| Take out existing window cill and replace | | | | | | |
| softwood | m | 0.55 | 8.80 | 9.52 | 2.75 | 21.07 |
| hardwood | m | 0.65 | 10.40 | 28.21 | 5.79 | 44.40 |

| | Unit | Labour hours | Labour cost £ | Materials £ | O & P £ | Total £ |
|---|---|---|---|---|---|---|
| **Take off existing external door and frame and replace with new** | | | | | | |
| flush door | | | | | | |
| softwood | nr | 2.20 | 35.20 | 110.09 | 21.79 | 167.08 |
| hardwood | nr | 2.50 | 40.00 | 393.03 | 64.95 | 497.98 |
| PVC-U | nr | 2.20 | 35.20 | 364.06 | 59.89 | 459.15 |
| panelled door | | | | | | |
| softwood | nr | 2.20 | 35.20 | 138.43 | 26.04 | 199.67 |
| hardwood | nr | 2.50 | 40.00 | 433.82 | 71.07 | 544.89 |
| PVC-U | nr | 2.20 | 35.20 | 407.66 | 66.43 | 509.29 |
| half glazed door | | | | | | |
| softwood | nr | 2.20 | 35.20 | 138.43 | 26.04 | 199.67 |
| hardwood | nr | 2.50 | 40.00 | 433.82 | 71.07 | 544.89 |
| PVC-U | nr | 2.20 | 35.20 | 407.66 | 66.43 | 509.29 |
| fully glazed door | | | | | | |
| softwood | nr | 2.80 | 44.80 | 111.18 | 23.40 | 179.38 |
| hardwood | nr | 3.20 | 51.20 | 410.93 | 69.32 | 531.45 |
| PVC-U | nr | 2.80 | 44.80 | 379.21 | 63.60 | 487.61 |
| fully glazed patio doors | | | | | | |
| softwood | pair | 3.80 | 60.80 | 250.85 | 46.75 | 358.40 |
| hardwood | pair | 4.30 | 68.80 | 576.61 | 96.81 | 742.22 |
| PVC-U | pair | 2.80 | 44.80 | 531.92 | 86.51 | 663.23 |
| **Take off and replace damaged ironmongery to softwood doors** | | | | | | |
| bolts | | | | | | |
| barrel | nr | 0.20 | 3.20 | 6.98 | 1.53 | 11.71 |
| flush | nr | 0.20 | 3.20 | 9.10 | 1.85 | 14.15 |
| tower | nr | 0.30 | 4.80 | 7.98 | 1.92 | 14.70 |
| butts | | | | | | |
| light | pair | 0.20 | 3.20 | 4.18 | 1.11 | 8.49 |
| medium | pair | 0.25 | 4.00 | 4.25 | 1.24 | 9.49 |
| heavy | pair | 0.30 | 4.80 | 5.35 | 1.52 | 11.67 |

|  | Unit | Labour hours | Labour cost £ | Materials £ | O & P £ | Total £ |
|---|---|---|---|---|---|---|
| locks |  |  |  |  |  |  |
| cupboard | nr | 0.30 | 4.80 | 8.24 | 1.96 | 15.00 |
| mortice dead lock | nr | 0.85 | 13.60 | 14.17 | 4.17 | 31.94 |
| rim lock | nr | 0.45 | 7.20 | 6.43 | 2.04 | 15.67 |
| cylinder | nr | 1.10 | 17.60 | 28.17 | 6.87 | 52.64 |

Take off and replace damaged
ironmongery to hardwood doors

|  | Unit | Labour hours | Labour cost £ | Materials £ | O & P £ | Total £ |
|---|---|---|---|---|---|---|
| bolts |  |  |  |  |  |  |
| barrel | nr | 0.30 | 4.80 | 6.98 | 1.77 | 13.55 |
| flush | nr | 0.30 | 4.80 | 9.10 | 2.09 | 15.99 |
| tower | nr | 0.40 | 6.40 | 7.98 | 2.16 | 16.54 |
| butts |  |  |  |  |  |  |
| light | pair | 0.30 | 4.80 | 4.19 | 1.35 | 10.34 |
| medium | pair | 0.35 | 5.60 | 4.25 | 1.48 | 11.33 |
| heavy | pair | 0.40 | 6.40 | 5.35 | 1.76 | 13.51 |
| locks |  |  |  |  |  |  |
| cupboard | nr | 0.30 | 4.80 | 8.24 | 1.96 | 15.00 |
| mortice dead lock | nr | 0.85 | 13.60 | 14.17 | 4.17 | 31.94 |
| rim lock | nr | 0.45 | 7.20 | 6.43 | 2.04 | 15.67 |
| cylinder | nr | 1.10 | 17.60 | 28.17 | 6.87 | 52.64 |

# Part Three

## APPROXIMATE ESTIMATING

Excavation and filling

Concrete work

Brickwork and blockwork

Masonry

Carpentry and joinery

Finishings

Plumbing and heating

Painting

Wallpapering

Drainage

| | Unit | Total £ |
|---|---|---|

## APPROXIMATE ESTIMATING

The following rates are based upon the unit rates in Part One of this book but have been rounded off for ease of use

## EXCAVATION AND FILLING

Excavate for trench including supporting sides, level and ram bottom, part return, fill and ram, part load into skip, width 600mm

| | Unit | Total £ |
|---|---|---|
| by hand, depth | | |
| 0.75m | m | 45.00 |
| 1.00m | m | 62.00 |
| 1.50m | m | 97.00 |
| 2.00m | m | 137.00 |
| | | |
| by machine, depth | | |
| 0.75m | m | 33.00 |
| 1.00m | m | 43.00 |
| 1.50m | m | 66.00 |
| 2.00m | m | 89.00 |

Filling in layers by hand over 250mm thick

| | Unit | Total £ |
|---|---|---|
| surplus excavated material | m3 | 20.00 |
| sand | m3 | 58.00 |
| hardcore | m3 | 44.00 |

Filling in layers by machine over 250mm thick

| | Unit | Total £ |
|---|---|---|
| surplus excavated material | m3 | 7.00 |
| sand | m3 | 45.00 |
| hardcore | m3 | 30.00 |

|  | Unit | Total £ |
|---|---|---|

## CONCRETE WORK

Ready mixed concrete 1:3:6
40mm aggregate in foundations,
size

| | | |
|---|---|---|
| 600 x 225mm | m | 17.00 |
| 750 x 225mm | m | 20.00 |

Ready mixed concrete 1:2:4
20mm aggregate, in beds including
mesh reinforcement and trowelling
smooth

| beds, thickness | | |
|---|---|---|
| 100mm | m2 | 19.00 |
| 150mm | m2 | 26.00 |
| 200mm | m2 | 33.00 |

Ready mixed concrete 1:2:4
20mm aggregate, in slabs including
mesh reinforcement and formwork
to soffit

| | | |
|---|---|---|
| 100mm | m2 | 64.00 |
| 150mm | m2 | 71.00 |
| 200mm | m2 | 78.00 |

Ready mixed concrete 1:2:4
20mm aggregate, in walls including
mesh reinforcement and formwork
both sides

| | | |
|---|---|---|
| 100mm | m2 | 119.00 |
| 150mm | m2 | 126.00 |
| 200mm | m2 | 128.00 |

| | Unit | Total £ |
|---|---|---|

## BRICKWORK AND BLOCKWORK

Cavity wall in cement mortar
in foundations including
forming cavity and wall ties
between outer leaf in common
bricks £140 per 1000 and
inner leaf of

| | Unit | Total |
|---|---|---|
| common brick £140 per 1000 | m2 | 116.00 |
| blockwork 100mm thick | m2 | 98.00 |

Cavity wall in cement mortar
in foundations including
forming cavity and wall ties
between outer leaf in common
bricks £200 per 1000 and
inner leaf of

| | Unit | Total |
|---|---|---|
| common brick £200 per 1000 | m2 | 128.00 |
| blockwork 100mm thick | m2 | 104.00 |

Cavity wall in cement mortar
in foundations including
forming cavity and wall ties
between outer leaf in engineering
bricks £250 per 1000 and
inner leaf of

| | Unit | Total |
|---|---|---|
| engineering bricks £250 per 1000 | m2 | 142.00 |
| blockwork 100mm thick | m2 | 111.00 |

Cavity wall in cement mortar
in foundations including forming
cavity and wall ties between outer
leaf in engineering bricks £350 per
1000 and inner leaf of

| | Unit | Total |
|---|---|---|
| engineering bricks £350 per 1000 | m2 | 160.00 |
| blockwork 100mm thick | m2 | 120.00 |

| | Unit | Total £ |
|---|---|---|
| **Cavity wall in gauged mortar including forming cavity and wall ties between outer leaf in facing bricks £250 per 1000 and inner leaf of** | | |
| common brick £140 per 1000 | m2 | 126.00 |
| common brick £200 per 1000 | m2 | 132.00 |
| blockwork 100mm thick | m2 | 108.00 |
| **Cavity wall in gauged mortar including forming cavity and wall ties between outer leaf in facing bricks £400 per 1000 and inner leaf of** | | |
| common brick £140 per 1000 | m2 | 136.00 |
| common brick £200 per 1000 | m2 | 142.00 |
| blockwork 100mm thick | m2 | 118.00 |
| **Cavity wall in gauged mortar including forming cavity and wall ties between outer leaf in facing bricks £500 per 1000 and inner leaf of** | | |
| common brick £140 per 1000 | m2 | 136.00 |
| common brick £200 per 1000 | m2 | 142.00 |
| blockwork 100mm thick | m2 | 118.00 |

## MASONRY

| | Unit | Total £ |
|---|---|---|
| **Cavity wall in gauged mortar including forming cavity and wall ties between outer leaf in irregular coursed rubble walling 300mm wide and inner leaf of** | | |
| common brick £140 per 1000 | m2 | 166.00 |
| common brick £200 per 1000 | m2 | 172.00 |

| | Unit | Total £ |
|---|---|---|

Cavity wall in gauged mortar including forming cavity and wall ties between outer leaf in coursed rubble walling 450mm wide and inner leaf of

| | | |
|---|---|---|
| common brick £140 per 1000 | m2 | 214.00 |
| common brick £200 per 1000 | m2 | 220.00 |

## CARPENTRY AND JOINERY

19mm thick butt-jointed flooring fixed to softwood joists size

| | | |
|---|---|---|
| 50 x 100mm | m2 | 35.00 |
| 50 x 125mm | m2 | 36.00 |
| 50 x 150mm | m2 | 38.00 |

25mm thick butt-jointed flooring fixed to softwood joists size

| | | |
|---|---|---|
| 50 x 100mm | m2 | 40.00 |
| 50 x 125mm | m2 | 41.00 |
| 50 x 150mm | m2 | 43.00 |

19mm thick tongued and grooved flooring fixed to softwood joists size

| | | |
|---|---|---|
| 50 x 100mm | m2 | 39.00 |
| 50 x 125mm | m2 | 40.00 |
| 50 x 150mm | m2 | 42.00 |

25mm thick tongued and grooved flooring fixed to softwood joists size

| | | |
|---|---|---|
| 50 x 100mm | m2 | 41.00 |
| 50 x 125mm | m2 | 42.00 |
| 50 x 150mm | m2 | 44.00 |

| | Unit | Total £ |
|---|---|---|
| **18mm thick chipboard butt-jointed flooring fixed to softwood joists size** | | |
| 50 x 100mm | m2 | 27.00 |
| 50 x 125mm | m2 | 28.00 |
| 50 x 150mm | m2 | 30.00 |
| **25mm thick chipboard butt-jointed flooring fixed to softwood joists size** | | |
| 50 x 100mm | m2 | 29.00 |
| 50 x 125mm | m2 | 30.00 |
| 50 x 150mm | m2 | 32.00 |
| **25mm thick chipboard tongued and grooved flooring fixed to softwood joists size** | | |
| 50 x 100mm | m2 | 31.00 |
| 50 x 125mm | m2 | 32.00 |
| 50 x 150mm | m2 | 34.00 |
| **Standard flush hardboard-faced internal door including lining and stops, architraves and fixing only hardware** | | |
| 35mm thick | | |
| 610 x 1981mm | nr | 162.00 |
| 686 x 1981mm | nr | 162.00 |
| 762 x 1981mm | nr | 162.00 |
| 40mm thick | | |
| 610 x 1981mm | nr | 164.00 |
| 686 x 1981mm | nr | 164.00 |
| 762 x 1981mm | nr | 164.00 |

| | Unit | Total £ |
|---|---|---|
| **Standard flush sapele-faced internal door including lining and stops, architraves and fixing only hardware** | | |
| 35mm thick | | |
|   610 x 1981mm | nr | 181.00 |
|   686 x 1981mm | nr | 181.00 |
|   762 x 1981mm | nr | 181.00 |
| 40mm thick | | |
|   610 x 1981mm | nr | 185.00 |
|   686 x 1981mm | nr | 185.00 |
|   762 x 1981mm | nr | 185.00 |
| **Standard flush hardboard-faced half-hour fire-check internal door including lining and stops, architraves and fixing only hardware** | | |
| 44mm thick | | |
|   686 x 1981mm | nr | 213.00 |
|   762 x 1981mm | nr | 214.00 |
|   726 x 2040mm | nr | 218.00 |
|   826 x 2040mm | nr | 219.00 |
| **Standard flush sapele-faced half-hour fire-check internal door including lining and stops, architraves and fixing only hardware** | | |
| 44mm thick | | |
|   686 x 1981mm | nr | 245.00 |
|   762 x 1981mm | nr | 248.00 |
|   726 x 2040mm | nr | 248.00 |
|   826 x 2040mm | nr | 253.00 |

|  | Unit | Total £ |
|---|---|---|

Standard flush hardboard-faced external door including frame, architraves and fixing only hardware

44mm thick

| | Unit | Total £ |
|---|---|---|
| 762 x 1981mm | nr | 220.00 |
| 838 x 1981mm | nr | 223.00 |

## FINISHINGS

One coat 'Universal' plaster to walls, plasterboard and skim to ceilings in rooms, storey height 2.4m

floor area

| | Unit | Total £ |
|---|---|---|
| 9m2 | nr | 425.00 |
| 12m2 | nr | 516.00 |
| 15m2 | nr | 631.00 |
| 18m2 | nr | 737.00 |
| 21m2 | nr | 819.00 |
| 24m2 | nr | 892.00 |
| 27m2 | nr | 993.00 |
| 30m2 | nr | 1,194.00 |

## PLUMBING AND HEATING

PVC-U rainwater goods

one-storey gable end

| | Unit | Total £ |
|---|---|---|
| terraced | nr | 283.00 |
| semi-detached | nr | 332.00 |
| detached | nr | 383.00 |

two-storey gable end

| | Unit | Total £ |
|---|---|---|
| terraced | nr | 344.00 |
| semi-detached | nr | 390.00 |
| detached | nr | 422.00 |

| | Unit | Total |
|---|---|---|
| | | £ |
| three-storey gable end | | |
| terraced | nr | 429.00 |
| semi-detached | nr | 455.00 |
| detached | nr | 501.00 |
| one-storey hipped end | | |
| terraced | nr | 279.00 |
| semi-detached | nr | 409.00 |
| detached | nr | 527.00 |
| two-storey hipped end | | |
| terraced | nr | 466.00 |
| semi-detached | nr | 492.00 |
| detached | nr | 610.00 |
| three-storey hipped end | | |
| terraced | nr | 483.00 |
| semi-detached | nr | 590.00 |
| detached | nr | 723.00 |
| Cast iron rainwater goods | | |
| one-storey gable end | | |
| terraced | nr | 743.00 |
| semi-detached | nr | 806.00 |
| detached | nr | 876.00 |
| two-storey gable end | | |
| terraced | nr | 967.00 |
| semi-detached | nr | 1,011.00 |
| detached | nr | 1,074.00 |
| three-storey gable end | | |
| terraced | nr | 1,129.00 |
| semi-detached | nr | 1,268.00 |
| detached | nr | 1,321.00 |

|  | Unit | Total £ |
|---|---|---|
| one-storey hipped end | | |
| terraced | nr | 704.00 |
| semi-detached | nr | 946.00 |
| detached | nr | 1,201.00 |
| two-storey hipped end | | |
| terraced | nr | 950.00 |
| semi-detached | nr | 1,238.00 |
| detached | nr | 1,467.00 |
| three-storey hipped end | | |
| terraced | nr | 1,092.00 |
| semi-detached | nr | 1,400.00 |
| detached | nr | 1,650.00 |
| Central heating | | |
| one-storey house overall size | | |
| (5 radiators) | nr | 2,561.00 |
| (6 radiators) | nr | 2,779.00 |
| (7 radiators) | nr | 2,959.00 |
| two-storey house overall size | | |
| (8 radiators) | nr | 3,488.00 |
| (9 radiators) | nr | 3,782.00 |
| (10 radiators) | nr | 4,207.00 |
| three-storey house overall size | | |
| (12 radiators) | nr | 5,842.00 |
| (14 radiators) | nr | 6,202.00 |
| (15 radiators) | nr | 6,420.00 |

## PAINTING

One mist coat and two coats of
emulsion paint to walls and ceilings
in rooms, storey height 2.4m

| floor area | | |
|---|---|---|
| 9m2 | nr | 261.00 |

| | Unit | Total £ |
|---|---|---|
| 12m2 | nr | 324.00 |
| 15m2 | nr | 396.00 |
| 18m2 | nr | 459.00 |
| 21m2 | nr | 513.00 |
| 24m2 | nr | 558.00 |
| 27m2 | nr | 621.00 |
| 30m2 | nr | 684.00 |

## WALLPAPERING

Prepare and hang wallpaper £5 per
roll to walls in rooms, storey
height 2.4m

floor area
| | Unit | Total £ |
|---|---|---|
| 9m2 | nr | 160.00 |
| 12m2 | nr | 192.00 |
| 15m2 | nr | 232.00 |
| 18m2 | nr | 264.00 |
| 21m2 | nr | 288.00 |
| 24m2 | nr | 304.00 |
| 27m2 | nr | 336.00 |
| 30m2 | nr | 365.00 |

Prepare and hang wallpaper £7 per
roll to walls in rooms, storey
height 2.4m

floor area
| | Unit | Total £ |
|---|---|---|
| 9m2 | nr | 180.00 |
| 12m2 | nr | 216.00 |
| 15m2 | nr | 261.00 |
| 18m2 | nr | 297.00 |
| 21m2 | nr | 324.00 |
| 24m2 | nr | 342.00 |
| 27m2 | nr | 378.00 |
| 30m2 | nr | 414.00 |

| | Unit | Total £ |
|---|---|---|

Prepare and hang wallpaper £9 per
roll to walls in rooms, storey
height 2.4m

floor area

| | | |
|---|---|---|
| 9m2 | nr | 240.00 |
| 12m2 | nr | 288.00 |
| 15m2 | nr | 348.00 |
| 18m2 | nr | 396.00 |
| 21m2 | nr | 432.00 |
| 24m2 | nr | 456.00 |
| 27m2 | nr | 504.00 |
| 30m2 | nr | 552.00 |

## DRAINAGE

Excavate trench by hand, lay
100mm diameter 'Hepsleve' pipe,
granular bed and surround, trench
depth

| | | |
|---|---|---|
| 0.50m | m | 45.00 |
| 0.75m | m | 55.00 |
| 1.00m | m | 65.00 |
| 1.25m | m | 90.00 |
| 1.50m | m | 105.00 |
| 1.75m | m | 120.00 |
| 2.00m | m | 140.00 |
| 2.25m | m | 157.00 |
| 2.50m | m | 177.00 |
| 2.75m | m | 195.00 |
| 3.00m | m | 220.00 |

Excavate trench by hand, lay
150mm diameter 'Hepsleve' pipe,
granular bed and surround, trench
depth

| | | |
|---|---|---|
| 0.50m | m | 65.00 |
| 0.75m | m | 75.00 |

|  | Unit | Total £ |
|---|---|---|
| 1.00m | m | 95.00 |
| 1.25m | m | 110.00 |
| 1.50m | m | 125.00 |
| 1.75m | m | 140.00 |
| 2.00m | m | 160.00 |
| 2.25m | m | 177.00 |
| 2.50m | m | 197.00 |
| 2.75m | m | 215.00 |
| 3.00m | m | 240.00 |

Excavate trench by machine, lay 100mm diameter 'Hepsleve' pipe, granular bed and surround, trench depth

| | | |
|---|---|---|
| 0.50m | m | 33.00 |
| 0.75m | m | 37.00 |
| 1.00m | m | 45.00 |
| 1.25m | m | 57.00 |
| 1.50m | m | 63.00 |
| 1.75m | m | 71.00 |
| 2.00m | m | 82.00 |
| 2.25m | m | 95.00 |
| 2.50m | m | 105.00 |
| 2.75m | m | 117.00 |
| 3.00m | m | 102.00 |

Excavate trench by machine, lay 100mm diameter 'Hepsleve' pipe, granular bed and surround, trench depth

| | | |
|---|---|---|
| 0.50m | m | 53.00 |
| 0.75m | m | 57.00 |
| 1.00m | m | 65.00 |
| 1.25m | m | 77.00 |
| 1.50m | m | 83.00 |
| 1.75m | m | 91.00 |
| 2.00m | m | 102.00 |
| 2.25m | m | 115.00 |

| | Unit | Total £ |
|---|---|---|
| 2.50m | m | 125.00 |
| 2.75m | m | 142.00 |
| 3.00m | m | 152.00 |

Manhole complete including hand
excavation, concrete base and
benching, engineering brickwork,
channels and cast iron cover,
depth

| | | |
|---|---|---|
| 1.00m | nr | 575.00 |
| 1.50m | nr | 778.00 |
| 2.00m | nr | 1006.00 |

Manhole complete including machine
excavation, concrete base and
benching, engineering brickwork,
channels and cast iron cover,
depth

| | | |
|---|---|---|
| 1.00m | nr | 556.00 |
| 1.50m | nr | 719.00 |
| 2.00m | nr | 874.00 |

# Part Four

## PLANT AND TOOL HIRE

Concrete and cutting equipment

Access and site equipment

Lifting and moving

Compaction

Breaking and demolition

Power tools

Welding and generators

Pumping equipment

| | 24 hours | Additional 24 hours | Week |
|---|---|---|---|
| | £ | £ | £ |

## PLANT AND TOOL HIRE

These selected rates are based on
average hire charges made by
hire firms in UK. Check your
local dealer for more information.
These prices exclude VAT.

## CONCRETE AND CUTTING EQUIPMENT

### Concrete mixers

| | 24 hours | Additional 24 hours | Week |
|---|---|---|---|
| Petrol, with stand | 16.00 | 8.00 | 32.00 |
| Electric, with stand | 14.00 | 7.00 | 28.00 |
| Bulk mixer | 36.00 | 18.00 | 72.00 |

### Vibrating pokers

| | 24 hours | Additional 24 hours | Week |
|---|---|---|---|
| Pokers | | | |
| petrol | 50.00 | 25.00 | 100.00 |
| electric | 38.00 | 19.00 | 76.00 |
| air poker, 25mm | 30.00 | 15.00 | 60.00 |
| air poker, 50mm | 32.00 | 16.00 | 64.00 |

### Power floats

| | 24 hours | Additional 24 hours | Week |
|---|---|---|---|
| Floats | | | |
| power float, petrol | 46.00 | 23.00 | 92.00 |

### Vibrating screeds

| | 24 hours | Additional 24 hours | Week |
|---|---|---|---|
| Screed units | | | |
| with 5m beam | 58.00 | 29.00 | 116.00 |
| roller screed unit | 108.00 | 54.00 | 216.00 |

|  | 24 hours | Additional 24 hours | Week |
|---|---|---|---|
|  | £ | £ | £ |

## Floor preparation units

| Floor saw, petrol | | | |
|---|---|---|---|
| 350mm | 64.00 | 32.00 | 128.00 |
| 450mm | 76.00 | 38.00 | 152.00 |
| Scabbler, hand held | 26.00 | 13.00 | 52.00 |
| Diamond concrete planer | 70.00 | 35.00 | 140.00 |
| Air needle gun | 30.00 | 15.00 | 60.00 |

## Disc cutters

| Cutters | | | |
|---|---|---|---|
| electric, 300mm | 24.00 | 12.00 | 48.00 |
| two stroke, 300mm | 28.00 | 14.00 | 56.00 |
| two stroke, 350mm | 30.00 | 15.00 | 60.00 |
| electric wall chasers | 48.00 | 24.00 | 96.00 |

## Block and slab splitters

| Splitters | | | |
|---|---|---|---|
| clay | 28.00 | 14.00 | 56.00 |
| block | 24.00 | 12.00 | 48.00 |
| slab | 36.00 | 18.00 | 72.00 |

## Circular saws

| Electric, 150mm | 18.00 | 9.00 | 36.00 |
|---|---|---|---|
| Electric, 230mm | 20.00 | 10.00 | 40.00 |

## ACCESS AND SITE EQUIPMENT

## Ladders

| Double ladder, alloy | | | |
|---|---|---|---|
| 4m | 16.00 | 8.00 | 32.00 |
| 6m | 24.00 | 12.00 | 48.00 |

| | 24 hours £ | Additional 24 hours £ | Week £ |
|---|---|---|---|
| Triple ladder, alloy | | | |
| 9m | 20.00 | 10.00 | 40.00 |
| Roof ladder | | | |
| 5m | 26.00 | 13.00 | 52.00 |
| Rope operated | | | |
| 11m | 40.00 | 20.00 | 80.00 |
| 13m | 46.00 | 23.00 | 92.00 |
| 16m | 54.00 | 27.00 | 108.00 |

## Props

| | 24 hours £ | Additional 24 hours £ | Week £ |
|---|---|---|---|
| Shoring props | | | |
| type 0 | 0.00 | 0.00 | 8.00 |
| type 1 | 0.00 | 0.00 | 8.00 |
| type 2 | 0.00 | 0.00 | 8.00 |
| type 3 | 0.00 | 0.00 | 8.00 |
| type 4 | 0.00 | 0.00 | 8.00 |

## Rubbish chutes

| | 24 hours £ | Additional 24 hours £ | Week £ |
|---|---|---|---|
| Chutes | | | |
| 1m section | 12.00 | 6.00 | 24.00 |
| funnel | 22.00 | 11.00 | 44.00 |
| hopper | 20.00 | 10.00 | 40.00 |

## Trestles and staging

| | 24 hours £ | Additional 24 hours £ | Week £ |
|---|---|---|---|
| Staging | | | |
| 2.4m | 18.00 | 9.00 | 36.00 |
| 3.6m | 20.00 | 10.00 | 40.00 |
| 4.8m | 26.00 | 13.00 | 52.00 |
| Painters' trestle | | | |
| 1.8m | 10.00 | 5.00 | 20.00 |

| | 24 hours | Additional 24 hours | Week |
|---|---|---|---|
| | £ | £ | £ |

**Alloy towers**

Single width, height

| | | | |
|---|---|---|---|
| 2.30m | 46.00 | 23.00 | 92.00 |
| 3.20m | 56.00 | 28.00 | 112.00 |
| 4.20m | 68.00 | 34.00 | 136.00 |
| 5.20m | 78.00 | 39.00 | 156.00 |
| 6.20m | 88.00 | 44.00 | 176.00 |
| 7.20m | 96.00 | 48.00 | 192.00 |
| 8.20m | 108.00 | 54.00 | 216.00 |
| 9.20m | 116.00 | 58.00 | 232.00 |
| 10.20m | 126.00 | 63.00 | 252.00 |

## LIFTING AND MOVING

| | | | |
|---|---|---|---|
| Paving stone lifter | 16.00 | 8.00 | 32.00 |
| Block transport cart | 80.00 | 40.00 | 160.00 |
| Plasterboard jack lift | 40.00 | 20.00 | 80.00 |
| Stair climber trolley | 16.00 | 8.00 | 32.00 |
| Rubble truck | 24.00 | 12.00 | 48.00 |

## COMPACTION

| | | | |
|---|---|---|---|
| Vibrating roller | 64.00 | 32.00 | 138.00 |

**Rammers**

| | | | |
|---|---|---|---|
| Two stroke | 48.00 | 24.00 | 96.00 |
| Four stroke | 50.00 | 25.00 | 100.00 |

## BREAKING AND DEMOLITION

**Breakers**

Hydraulic

| | | | |
|---|---|---|---|
| diesel | 74.00 | 37.00 | 128.00 |
| petrol | 80.00 | 40.00 | 160.00 |
| medium duty, electric | 24.00 | 12.00 | 48.00 |
| heavy duty, electric | 48.00 | 24.00 | 96.00 |
| Air breaker, medium | 34.00 | 17.00 | 68.00 |
| Air breaker, heavy | 36.00 | 18.00 | 72.00 |

| | 24 hours £ | Additional 24 hours £ | Week £ |
|---|---|---|---|
| **POWER TOOLS** | | | |
| **Drills** | | | |
| Cordless drill | 20.00 | 10.00 | 40.00 |
| Right angle drill | 22.00 | 11.00 | 24.00 |
| Combi hammer | | | |
| light duty | 22.00 | 11.00 | 44.00 |
| heavy duty | 28.00 | 14.00 | 56.00 |
| Hammer drill | | | |
| lightweight | 16.00 | 8.00 | 32.00 |
| heavyweight | 20.00 | 10.00 | 40.00 |
| Lightweight diamond driller kit | 48.00 | 24.00 | 96.00 |
| Four speed rotary drill | 30.00 | 15.00 | 60.00 |
| **Grinders** | | | |
| Angle grinder | | | |
| 100mm | 16.00 | 8.00 | 32.00 |
| 125mm | 16.00 | 8.00 | 32.00 |
| 230mm | 16.00 | 8.00 | 32.00 |
| 300mm | 30.00 | 15.00 | 60.00 |
| **Saws** | | | |
| Reciprocating saw | | | |
| standard | 24.00 | 12.00 | 48.00 |
| heavy duty | 28.00 | 14.00 | 56.00 |
| Circular saw | | | |
| 150mm | 18.00 | 9.00 | 36.00 |
| 230mm | 20.00 | 10.00 | 40.00 |

| | 24 hours | Additional 24 hours | Week |
|---|---|---|---|
| | £ | £ | £ |

**Woodworking**

| | | | |
|---|---|---|---|
| Router | 20.00 | 10.00 | 40.00 |
| Worktop jig | 18.00 | 9.00 | 36.00 |

**Fixing equipment**

| | | | |
|---|---|---|---|
| Cordless nailing gun | 28.00 | 14.00 | 56.00 |
| Cartridge hammer | 28.00 | 14.00 | 56.00 |
| Electric screwdriver | 16.00 | 8.00 | 32.00 |

**Sanders**

| | | | |
|---|---|---|---|
| Floor | 38.00 | 19.00 | 76.00 |
| Floor edger | 30.00 | 15.00 | 60.00 |
| Orbital | 20.00 | 10.00 | 40.00 |

**WELDING AND GENERATORS**

**Generating**

Generators

| | | | |
|---|---|---|---|
| petrol, 3KVA | 28.00 | 14.00 | 56.00 |
| petrol, 5KVA | 40.00 | 20.00 | 80.00 |
| silenced, 6KVA | 90.00 | 45.00 | 180.00 |
| silenced, 20KVA | 220.00 | 110.00 | 440.00 |

**PUMPING EQUIPMENT**

**Pumps**

Submersible

| | | | |
|---|---|---|---|
| 50mm | 40.00 | 20.00 | 80.00 |
| Puddle pump, electric | 46.00 | 23.00 | 92.00 |
| Diaphragm pump, 50mm | 60.00 | 30.00 | 120.00 |

# Part Five

## GENERAL CONSTRUCTION DATA

## GENERAL CONSTRUCTION DATA

### The metric system

Linear

| | | |
|---|---|---|
| 1 centimetre (cm) | = | 10 millimetres (mm) |
| 1 decimetre  (dm) | = | 10 centimetres (cm) |
| 1 metre      (m) | = | 10 decimetres (dm) |
| 1 kilometre  (km) | = | 1000 metres (m) |

Area

| | | |
|---|---|---|
| 100 sq millimetres | = | 1 sq centimetre |
| 100 sq centimetres | = | 1 sq decimetre |
| 100 sq decimetres | = | 1 sq metre |
| 1000 sq metres | = | 1 hectare |

Capacity

| | | |
|---|---|---|
| 1 millilitre (ml) | = | 1 cubic centimetre (cm3) |
| 1 centilitre (cl) | = | 10 millilitres (ml) |
| 1 decilitre (dl) | = | 10 centilitres (cl) |
| 1 litre (l) | = | 10 decilitres (dl) |

Weight

| | | |
|---|---|---|
| 1 centigram (cg) | = | 10 milligrams (mg) |
| 1 decigram (dg) | = | 10 centigrams (mcg) |
| 1 gram (g) | = | 10 decigrams (dg) |
| 1 decagram (dag) | = | 10 grams (g) |
| 1 hectogram (hg) | = | 10 decagrams (dag) |

### Conversion equivalents (imperial/metric)

Length

| | | |
|---|---|---|
| 1 inch | = | 25.4 mm |
| 1 foot | = | 304.8 mm |
| 1 yard | = | 914.4 mm |
| 1 yard | = | 0.9144 m |
| 1 mile | = | 1609.34 m |

Area

| | | |
|---|---|---|
| 1 sq inch | = | 645.16 sq mm |
| 1 sq ft | = | 0.092903 sq m |
| 1 sq yard | = | 0.8361 sq m |
| 1 acre | = | 4840 sq yards |
| 1 acre | = | 2.471 hectares |

Liquid

| | | |
|---|---|---|
| 1 lb water | = | 0.454 litres |
| 1 pint | = | 0.568 litres |
| 1 gallon | = | 4.546 litres |

Horse-power

| | | |
|---|---|---|
| 1 hp | = | 746 watts |
| 1 hp | = | 0.746 kW |
| 1 hp | = | 33,000 ft.lb/min |

Weight

| | | |
|---|---|---|
| 1 lb | = | 0.4536 kg |
| 1 cwt | = | 50.8 kg |
| 1 ton | = | 1016.1 kg |

## Conversion equivalents (metric/imperial)

Length

| | | |
|---|---|---|
| 1 mm | = | 0.03937 inches |
| 1 centimetre | = | 0.3937 inches |
| 1 metre | = | 1.094 yards |
| 1 metre | = | 3.282 ft |
| 1 kilometre | = | 0.621373 miles |

Area

| | | |
|---|---|---|
| 1 sq millimetre | = | 0.00155 sq in |
| 1 sq metre | = | 10.764 sq ft |
| 1 sq metre | = | 1.196 sq yards |
| 1 acre | = | 4046.86 sq m |
| 1 hectare | = | 0.404686 acres |

Liquid

| | | |
|---|---|---|
| 1 litre | = | 2.202 lbs |
| 1 litre | = | 1.76 pints |
| 1 litre | = | 0.22 gallons |

Horse-power

|  |  |  |
|---|---|---|
| 1 watt | = | 0.00134 hp |
| 1 kw | = | 134 hp |
| 1 hp | = | 0759 kg m/s |

Weight

|  |  |  |
|---|---|---|
| 1 kg | = | 2.205 lbs |
| 1 kg | = | 0.01968 cwt |
| 1 kg | = | 0.000984 ton |

## Temperature equivalents

In order to convert Fahrenheit to Celsius deduct 32 and multiply by 5/9.
To convert Celsius to Fahrenheit multiply by 9/5 and add 32.

| Fahrenheit | Celsius |
|---|---|
| 230 | 110.0 |
| 220 | 104.4 |
| 210 | 98.9 |
| 200 | 93.3 |
| 190 | 87.8 |
| 180 | 82.2 |
| 170 | 76.7 |
| 160 | 71.1 |
| 150 | 65.6 |
| 140 | 60.0 |
| 130 | 54.4 |
| 120 | 48.9 |
| 110 | 43.3 |
| 100 | 37.8 |
| 90 | 32.2 |
| 80 | 26.7 |
| 70 | 21.1 |
| 60 | 15.6 |
| 50 | 10.0 |
| 40 | 4.4 |
| 30 | -1.1 |
| 20 | -6.7 |
| 10 | -12.2 |
| 0 | -17.8 |

**Areas and volumes**

| Figure | Area | Perimeter |
|---|---|---|
| Rectangle | Length × breadth | Sum of sides |
| Triangle | Base × half of perpendicular height | Sum of sides |
| Quadrilateral | Sum of areas of contained triangles | Sum of sides |
| Trapezoidal | Sum of areas of contained triangles | Sum of sides |
| Trapezium | Half of sum of parallel sides × perpendicular height | Sum of sides |
| Parallelogram | Base × perpendicular height | Sum of sides |
| Regular polygon | Half sum of sides × half internal diameter | Sum of sides |
| Circle | pi × radius² | pi × diameter or pi × 2 × radius |

| Figure | Surface area | Volume |
|---|---|---|
| Cylinder | pi × 2 × radius × length (curved surface only) | pi × radius² × length |
| Sphere | pi × diameter² | 1.33 × pi × radius³ |

| Weights of materials | kg/m3 | kg/m2 | kg/m |
|---|---|---|---|
| Aggregate, coarse | 1,500 | | |
| Ashes | 800 | | |
| Ballast | 600 | | |
| Blockboard, standard | 940-1000 | | |
| Blockboard, tempered | 940-1060 | | |
| Blocks, natural aggregate | | | |
| 75mm | | 160.00 | |
| 100mm | | 215.00 | |
| 140mm | | 300.00 | |

| Weights of materials | kg/m3 | kg/m2 | kg/m |
|---|---|---|---|
| Blocks, lighweight aggregate | | | |
| 75mm | | 60.00 | |
| 100mm | | 80.00 | |
| 140mm | | 112.00 | |
| Bricks, Fletton | | 1,820.00 | |
| Bricks, engineering | | 2,250.00 | |
| Bricks, concrete | | 1,850.00 | |
| Brickwork, 112.5mm | | 220.00 | |
| Brickwork, 215mm | | 465.00 | |
| Brickwork, 327.5mm | | 710.00 | |
| Cement | 1,440 | | |
| Chalk | 2,240 | | |
| Chipboard, standard | 650-750 | | |
| Chipboard, flooring | 680-800 | | |
| Clay | 1,800 | | |
| Concrete | 2,450 | | |
| Copper pipes, table X | | | |
| 6mm | | | 0.091 |
| 8mm | | | 0.125 |
| 10mm | | | 0.158 |
| 12mm | | | 0.191 |
| 15mm | | | 0.280 |
| 18mm | | | 0.385 |
| 22mm | | | 0.531 |
| 28mm | | | 0.681 |
| 35mm | | | 1.133 |
| 42mm | | | 1.368 |
| 54mm | | | 1.769 |
| Copper pipes, table Y | | | |
| 6mm | | | 0.117 |
| 8mm | | | 0.162 |
| 10mm | | | 0.206 |
| 12mm | | | 0.251 |
| 15mm | | | 0.392 |
| 18mm | | | 0.476 |
| 22mm | | | 0.697 |
| 28mm | | | 0.899 |
| 35mm | | | 1.409 |
| 42mm | | | 1.700 |
| 54mm | | | 2.905 |

| Weights of materials | kg/m3 | kg/m2 | kg/m |
|---|---|---|---|
| Copper pipes, table Z | | | |
| 6mm | | | 0.077 |
| 8mm | | | 0.105 |
| 10mm | | | 0.133 |
| 12mm | | | 0.161 |
| 15mm | | | 0.203 |
| 18mm | | | 0.292 |
| 22mm | | | 0.359 |
| 28mm | | | 0.459 |
| 35mm | | | 0.670 |
| 42mm | | | 0.922 |
| 54mm | | | 1.334 |
| Flint | 2,550 | | |
| Gravel | 1,750 | | |
| Hardcore | 1,900 | | |
| Hoggin | 1,750 | | |
| Glass, clear sheet | | | |
| 3mm | | 7.50 | |
| 4mm | | 10.00 | |
| 5mm | | 12.50 | |
| 6mm | | 15.00 | |
| 10mm | | 25.00 | |
| 12mm | | 30.00 | |
| 15mm | | 37.50 | |
| 19mm | | 47.50 | |
| 25mm | | 63.50 | |
| Glass, float | | | |
| 3mm | | 7.50 | |
| 4mm | | 10.00 | |
| 5mm | | 12.50 | |
| 6mm | | 15.00 | |
| Glass, patterned | | | |
| 3mm | | 6.00 | |
| 4mm | | 7.50 | |
| 5mm | | 9.50 | |
| 6mm | | 11.50 | |
| 10mm | | 21.50 | |
| Laminboard | 500-700 | | |
| Lime, ground | 750 | | |

| Weights of materials | kg/m3 | kg/m2 | kg/m |
|---|---|---|---|
| Mild steel flat bars | | | |
| 25 × 9.53mm | | | 1.910 |
| 38 × 9.53mm | | | 2.840 |
| 50 × 12.70 | | | 5.060 |
| 50 × 19.00 | | | 7.590 |
| Mild steel round bars | | | |
| 6mm | | | 0.222 |
| 8mm | | | 0.395 |
| 10mm | | | 0.616 |
| 12mm | | | 0.888 |
| 16mm | | | 1.579 |
| 20mm | | | 2.466 |
| 25mm | | | 3.854 |
| 32mm | | | 6.313 |
| 40mm | | | 9.864 |
| 50mm | | | 15.413 |
| Mild steel square bars | | | |
| 6mm | | | 0.283 |
| 8mm | | | 0.503 |
| 10mm | | | 0.784 |
| 12mm | | | 1.131 |
| 16mm | | | 2.010 |
| 20mm | | | 3.139 |
| 25mm | | | 4.905 |
| 32mm | | | 8.035 |
| 40mm | | | 12.554 |
| 50mm | | | 19.617 |
| Plaster | | | |
| Carlite browning.11mm thick | | 7.80 | |
| Carlite tough coat, 11mm thick | | 7.80 | |
| Carlite bonding, 8mm thick | | 7.10 | |
| Carlite bonding, 11mm thick | | 9.80 | |
| Thistle hardwall, 11mm thick | | 8.80 | |
| Thistle dri-coat, 11mm thick | | 8.30 | |
| Thistle renovating, 11mm thick | | 8.80 | |
| Sand | 1,600 | | |
| Screed, 12.5mm thick | | 29.00 | |
| Stone, Bath | 2,200 | | |
| Stone, crushed | 1,350 | | |
| Stone, Darley Dale | 2,400 | | |

| Weights of materials | kg/m3 | kg/m2 | kg/m |
|---|---|---|---|
| Stone, natural | 2,400 | | |
| Stone, Portland | 2,200 | | |
| Stone, reconstructed | 2,250 | | |
| Stone, York | 2,400 | | |
| Terrazzo, 25mm thick | | 45.50 | |
| Timber | | | |
| Ash | 800 | | |
| Baltic Spruce | 480 | | |
| Beech | 815 | | |
| Birch | 720 | | |
| Box× | 960 | | |
| Cedar | 480 | | |
| Ebony | 1,215 | | |
| Elm | 625 | | |
| Greenheart | 960 | | |
| Jarrah | 815 | | |
| Maple | 750 | | |
| Pine, Pitchpine | 800 | | |
| Pine, Red Deal | 575 | | |
| Pine, Yellow Deal | 530 | | |
| Sycamore | 530 | | |
| Teak, African | 960 | | |
| Teak, Indian | 655 | | |
| Walnut | 495 | | |
| Top soil | 1,000 | | |
| Water | 950 | | |
| Woodblock flooring | | | |
| softwood | | 12.70 | |
| hardwood | | 17.60 | |
| Zinc sheeting | | 4.6 | |

## EXCAVATION AND FILLING

### Shrinkage of deposited material

| | |
|---|---|
| Clay | -10% |
| Gravel | -7.50% |
| Sandy soil | -12.50% |

**Bulking excavated material**

| | |
|---|---|
| Clay | 40% |
| Gravel | 25% |
| Sand | 20% |

| **Typical fuel comsumption for plant** | **Engine size kW** | **Litres per hour** |
|---|---|---|
| Compressors up to | 20 | 4.00 |
| | 30 | 6.50 |
| | 40 | 8.20 |
| | 50 | 9.00 |
| | 75 | 16.00 |
| | 100 | 20.00 |
| | 125 | 25.00 |
| | 150 | 30.00 |
| Concrete mixers up to | | |
| | 5 | 1.00 |
| | 10 | 2.40 |
| | 15 | 3.80 |
| | 20 | 5.00 |
| Dumpers | 5 | 1.30 |
| | 7 | 2.00 |
| | 10 | 3.00 |
| | 15 | 4.00 |
| | 20 | 4.90 |
| | 30 | 7.00 |
| | 50 | 12.00 |
| Excavators | 10 | 2.50 |
| | 20 | 4.50 |
| | 40 | 9.00 |
| | 60 | 13.00 |
| | 80 | 17.00 |
| Pumps | 5 | 1.10 |
| | 10 | 2.10 |
| | 15 | 3.20 |
| | 20 | 4.20 |
| | 25 | 5.50 |

## CONCRETE WORK

| Concrete mixes | Mix | Cement t | Sand m3 | Aggregate m3 | Water litres |
|---|---|---|---|---|---|
| | 1:1:2 | 0.50 | 0.45 | 0.70 | 208.00 |
| | 1:1.5:3 | 0.37 | 0.50 | 0.80 | 185.00 |
| | 1:2:4 | 0.30 | 0.54 | 0.85 | 175.00 |
| | 1:2.5:5 | 0.25 | 0.55 | 0.85 | 166.00 |
| | 1:3:6 | 0.22 | 0.55 | 0.85 | 160.00 |

## BRICKWORK AND BLOCKWORK

### Bricks per m2 (brick size 215 × 103.5 × 65mm)

Half brick wall
| | |
|---|---|
| stretcher bond | 59 |
| English bond | 89 |
| English garden wall bond | 74 |
| Flemish bond | 79 |

One brick wall
| | |
|---|---|
| English bond | 118 |
| Flemish bond | 118 |

One and a half brick wall
| | |
|---|---|
| English bond | 178 |
| Flemish bond | 178 |

Two brick wall
| | |
|---|---|
| English bond | 238 |
| Flemish bond | 238 |

Metric modular bricks

200 × 100 × 75mm
| | |
|---|---|
| 90mm thick | 133 |
| 190mm thick | 200 |

200 × 100 × 100mm

| | |
|---|---|
| 90mm thick | 50 |
| 190mm thick | 100 |
| 290mm thick | 150 |

300 × 100 × 75mm

| | |
|---|---|
| 90mm thick | 44 |

300 × 100 × 100mm

| | |
|---|---|
| 90mm thick | 50 |

Blocks per m2 (block size 414 × 215mm)

| | |
|---|---|
| 60mm thick | 9.9 |
| 75mm thick | 9.9 |
| 100mm thick | 9.9 |
| 140mm thick | 9.9 |
| 190mm thick | 9.9 |
| 215mm thick | 9.9 |

| **Mortar per m2** | **Wirecut** m3 | **1 Frog** m3 | **2 Frogs** m3 |
|---|---|---|---|
| Brick size 215 × 103.5 × 65mm | | | |
| Half brick wall | 0.017 | 0.024 | 0.031 |
| One brick wall | 0.045 | 0.059 | 0.073 |
| One and a half brick wall | 0.072 | 0.093 | 0.114 |
| Two brick wall | 0.101 | 0.128 | 0.155 |

| Brick size 200 × 100 × 75mm | **Solid** m3 | | **Perforated** m3 |
|---|---|---|---|
| 90mm thick | 0.016 | | 0.019 |
| 190mm thick | 0.042 | | 0.048 |
| 290mm thick | 0.068 | | 0.078 |

| Brick size 200 × 100 × 100mm | | | |
|---|---|---|---|
| 90mm thick | 0.013 | | 0.016 |
| 190mm thick | 0.036 | | 0.041 |
| 290mm thick | 0.059 | | 0.067 |

|  | Solid<br>m3 | Perforated<br>m3 |
|---|---|---|
| **Brick size 200 × 100 × 100mm** |  |  |
| 90mm thick | 0.015 | 0.018 |
| **Block size 440 × 215mm** |  |  |
| 60mm thick | 0.004 |  |
| 75mm thick | 0.005 |  |
| 100mm thick | 0.006 |  |
| 140mm thick | 0.007 |  |
| 190mm thick | 0.008 |  |
| 215mm thick | 0.009 |  |

## MASONRY

| **Mortar per m2 of random rubble walling** | m3 |
|---|---|
| 300mm thick wall | 0.120 |
| 450mm thick wall | 0.160 |
| 550mm thick wall | 0.120 |

## CARPENTRY AND JOINERY

| **Length of boarding required** | m/m2 |
|---|---|
| Board width, 75mm | 13.33 |
| Board width, 100mm | 10.00 |
| Board width, 125mm | 8.00 |
| Board width, 150mm | 6.67 |
| Board width, 175mm | 5.71 |
| Board width, 200mm | 5.00 |

| **ROOFING** | Lap<br>mm | Gauge<br>mm | Nr/m2 | Battens<br>m/m2 |
|---|---|---|---|---|
| Clay/concrete tiles |  |  |  |  |
| 267 × 165mm | 65 | 100 | 60.00 | 10.00 |
|  | 65 | 98 | 64.00 | 10.50 |
|  | 65 | 90 | 68.00 | 11.30 |

|  | Lap mm | Gauge mm | Nr/m2 | Battens m/m2 |
|---|---|---|---|---|
| 387 × 230mm | 75 | 300 | 16.00 | 3.20 |
|  | 100 | 280 | 17.40 | 3.50 |
| 420 × 330mm | 75 | 340 | 10.00 | 2.90 |
|  | 100 | 320 | 10.74 | 3.10 |
| **Fibre slates** |  |  |  |  |
| 500 × 250mm | 90 | 205 | 19.50 | 4.85 |
|  | 80 | 210 | 19.10 | 4.76 |
|  | 70 | 215 | 18.60 | 4.65 |
| 600 × 300mm | 105 | 250 | 13.60 | 4.04 |
|  | 100 | 250 | 13.40 | 4.00 |
|  | 90 | 255 | 13.10 | 3.92 |
|  | 80 | 260 | 12.90 | 3.85 |
|  | 70 | 263 | 12.70 | 3.77 |
| 400 × 200mm | 70 | 165 | 30.00 | 6.06 |
|  | 75 | 162 | 30.90 | 6.17 |
|  | 90 | 155 | 32.30 | 6.45 |
| 500 × 250mm | 70 | 215 | 18.60 | 4.65 |
|  | 75 | 212 | 18.90 | 4.72 |
|  | 90 | 205 | 19.50 | 4.88 |
|  | 100 | 200 | 20.00 | 5.00 |
|  | 110 | 195 | 20.50 | 5.13 |
| 600 × 300mm | 100 | 250 | 13.4 | 4.00 |
|  | 110 | 245 | 13.60 | 4.08 |
| **Natural slates** |  |  |  |  |
| 405 × 205mm | 75 | 165 | 29.59 | 8.70 |
| 405 × 255mm | 75 | 165 | 23.75 | 6.06 |
| 405 × 305mm | 75 | 165 | 19.00 | 5.00 |
| 460 × 230mm | 75 | 195 | 23.00 | 6.00 |
| 460 × 255mm | 75 | 195 | 20.37 | 5.20 |
| 460 × 305mm | 75 | 195 | 17.00 | 5.00 |
| 510 × 255mm | 75 | 220 | 18.02 | 4.60 |
| 510 × 305mm | 75 | 220 | 15.00 | 4.00 |

|  | Lap mm | Gauge mm | Nr/m2 | Battens m/m2 |
|---|---|---|---|---|
| 560 × 280mm | 75 | 240 | 14.81 | 4.12 |
| 560 × 280mm | 75 | 240 | 14.00 | 4.00 |
| 610 × 305mm | 75 | 265 | 12.27 | 3.74 |

Reconstructed stone slates

|  | Lap mm | Gauge mm | Nr/m2 | Battens m/m2 |
|---|---|---|---|---|
| 380 × 250mm | 75 | 150 | 16.00 | 3.20 |

## PLASTERING AND TILING

### Plaster coverage

|  | m2 per 1000kg |
|---|---|
| Carlite browning, 11mm thick | 135-155 |
| Carlite tough coat, 11mm thick | 135-150 |
| Carlite bonding, 11mm thick | 100-115 |
| Thistle hardwall, 11mm thick | 115-130 |
| Thistle dri-coat, 11mm thick | 135-135 |
| Thistle renovating, 11mm thick | 115-125 |

### Tile coverage

|  | Nr |
|---|---|
| 152 × 152mm | 43.27 |
| 200 × 200mm | 25.00 |

## PLUMBING AND HEATING

### Roof drainage

|  | Area m2 | Pipe mm | Gutter mm |
|---|---|---|---|
| One end outlet | 15 | 50 | 75 |
|  | 38 | 68 | 100 |
|  | 100 | 110 | 150 |
| Centre outlet | 30 | 50 | 75 |
|  | 75 | 68 | 100 |
|  | 200 | 110 | 150 |

## PAINTING AND WALLPAPERING

| Average coverage of paints m2 per litre | Timber | Plastered surfaces | Brickwork |
|---|---|---|---|
| Primer | 10-12 | 9-11 | 5-7 |
| Undercoat | 10-12 | 11-14 | 6-8 |
| Gloss | 11-14 | 11-14 | 6-8 |
| Emulsion | 10-12 | 12-15 | 6-10 |

| Wallpaper coverage per roll | Rolls nr | Wall height m | Room perimeter m |
|---|---|---|---|
| | 4 | 2.50 | 8 |
| | 5 | 2.50 | 9 |
| | 5 | 2.50 | 10 |
| | 6 | 2.50 | 11 |
| | 6 | 2.50 | 12 |
| | 7 | 2.50 | 13 |
| | 7 | 2.50 | 14 |
| | 8 | 2.50 | 15 |
| | 8 | 2.50 | 16 |
| | 8 | 2.50 | 17 |
| | 9 | 2.50 | 18 |
| | 10 | 2.50 | 19 |
| | 10 | 2.50 | 20 |
| | 10 | 2.50 | 21 |
| | 11 | 2.50 | 22 |
| | 11 | 2.50 | 23 |
| | 12 | 2.50 | 24 |
| | 13 | 2.50 | 25 |
| | 13 | 2.50 | 26 |
| | 14 | 2.50 | 27 |
| | 5 | 2.80 | 8 |
| | 5 | 2.80 | 9 |
| | 5 | 2.80 | 10 |
| | 7 | 2.80 | 11 |
| | 7 | 2.80 | 12 |
| | 7 | 2.80 | 13 |
| | 8 | 2.80 | 14 |
| | 8 | 2.80 | 15 |

| Rolls nr | Wall height m | Room perimeter m |
|---|---|---|
| 9 | 2.8 | 16 |
| 10 | 2.8 | 17 |
| 10 | 2.8 | 18 |
| 11 | 2.8 | 19 |
| 11 | 2.8 | 20 |
| 12 | 2.8 | 21 |
| 13 | 2.8 | 22 |
| 13 | 2.8 | 23 |
| 14 | 2.8 | 24 |
| 14 | 2.8 | 25 |
| 15 | 2.8 | 26 |
| 15 | 2.8 | 27 |

## EXTERNAL WORKS

### Blocks/slabs per m2

| | nr/m2 |
|---|---|
| 200 × 100mm | 50.00 |
| 450 × 450mm | 4.93 |
| 600 × 450mm | 3.70 |
| 600 × 600mm | 2.79 |
| 600 × 750mm | 2.22 |
| 600 × 900mm | 1.85 |

### Drainage trench widths

| | Under 1.5m deep mm | Over 1.5m deep mm |
|---|---|---|
| Pipe diameter 100mm | 450 | 600 |
| Pipe diameter 150mm | 500 | 650 |
| Pipe diameter 225mm | 600 | 750 |
| Pipe diameter 300mm | 650 | 800 |

| Volumes of filling for pipe beds (m3 per m) | 50mm thick | 100mm thick | 150mm thick |
|---|---|---|---|
| Pipe diameter 100mm | 0.023 | 0.045 | 0.068 |
| Pipe diameter 150mm | 0.026 | 0.053 | 0.079 |
| Pipe diameter 225mm | 0.030 | 0.060 | 0.090 |
| Pipe diameter 300mm | 0.038 | 0.075 | 0.113 |

**Volumes of filling for pipe
bed and haunching
(m3 per m)**

| | m3 |
|---|---|
| Pipe diameter 100mm | 0.117 |
| Pipe diameter 150mm | 0.152 |
| Pipe diameter 225mm | 0.195 |
| Pipe diameter 300mm | 0.279 |

**Volumes of filling for pipe
bed and surround
(m3 per m)**

| Pipe diameter 100mm | 0.185 |
|---|---|
| Pipe diameter 150mm | 0.231 |
| Pipe diameter 225mm | 0.285 |
| Pipe diameter 300mm | 0.391 |

# Part Six

## BUSINESS MATTERS

Starting a business

Running a business

Taxation

# STARTING A BUSINESS

Most small businesses come into being for one of two reasons – ambition or desperation! A person with genuine ambition for commercial success will never be completely satisfied until he has become self-employed and started his own business. But many successful businesses have been started because the proprietor was forced into this course of action because of redundancy.

Before giving up his job, the would-be businessman should consider carefully whether he has the required skills and the temperament to survive in the highly competitive self-employed market. Before commencing in business it is essential to assess the commercial viability of the intended business because it is pointless to finance a business that is not going to be commercially viable.

In the early stages it is important to make decisions such as: What exactly is the product being sold? What is the market view of that product? What steps are required before the developed product is first sold and where are those sales coming from?

As much information as possible should be obtained on how to run a business before taking the plunge. Sales targets should be set and it should be clearly established how those important first sales are obtained. Above all, do not underestimate the amount of time required to establish and finance a new business venture.

Whatever the size of the business it is important that you put in writing exactly what you are trying to do. This means preparing a business plan that will not only assist in establishing your business aims but is essential if you need to raise finance. The contents of a typical business plan are set out later. It is important to realise that you are not on your own and there are many contacts and advising agencies that can be of assistance.

## Potential customers and trade contacts

Many persons intending to start a business in the construction industry will have already had experience as employees. Use all contacts to check the market, establish the sort of work that is available and the current charge-out rates.

In the domestic market, check on the competition for prices and services provided. Study advertisements for your kind of work and try to get firm promises of work before the start-up date.

## Testing the market

Talk to as many traders as possible operating in the same field. Identify if the market is in the industrial, commercial, local government or in the domestic field. Talk to prospective customers and clients and consider how you can improve on what is being offered in terms of price, quality, speed, convenience, reliability and back-up service.

## Business links

There is no shortage of information about the many aspects of starting and running your own business. Finance, marketing, legal requirements, developing your business idea and taxation matters are all the subject of a mountain of books, pamphlets, guides and courses so it should not be necessary to pay out a lot of money for this information. Indeed, the likelihood is that the aspiring businessman will be overwhelmed with information and will need professional guidance to reduce the risk of wasting time on studying unnecessary subjects.

Business Links are now well established and provide a good place to start for both information and advice. These organisations provide a 'one-stop-shop' for advice and assistance to owner-managed businesses. They will often replace the need to contact Training and Enterprise Councils (TECs) and many of the other official organisations listed below.

Point of contact: telephone directory for address.

## Training and Enterprise Councils (TECs)

TECs are comprised of a board of directors drawn from the top men in local industry, commerce, education, trade unions etc., who, together with their staff and experienced business counsellors, assist both new and established concerns in all aspects of running a business. This takes the form of across-the-table advice and also hands-on assistance in management, marketing and finance if required. There are also training courses and seminars available in most areas together with the possibility of grants in some areas.

Point of contact: local Jobcentre or Citizens' Advice Bureau.

**Banks**

Approach banks for information about the business accounts and financial services that are available. Your local Business Link can advise on how best to find a suitable bank manager and inform you as to what the bank will require.

Shop around several banks and branches if you are not satisfied at first because managers vary widely in their views on what is a viable business proposition. Remember, most banks have useful free information packs to help business start-up.

Point of contact: local bank manager.

**HM Revenue and Customs**

Make a preliminary visit to the local tax office enquiry counter for their publications on income tax and national insurance contributions.

| | |
|---|---|
| SA/Bk 3 | Self assessment. A guide to keeping records for the self employed |
| IR 15(CIS) | Construction Industry Tax Deduction Scheme |
| CWF1 | Starting your own business, |
| IR 40(CIS) | Conditions for Getting a Sub-Contractor's Tax Certificate |
| NE1 | PAYE for Employers (if you employ someone) |
| NE3 | PAYE for new and small Employers |
| IR 56/N139 | Employed or Self-Employed. A guide for tax and National Insurance |
| CA02 | National Insurance contributions for self employed people with small earnings. |

Remember, the onus is on the taxpayer, within three months, to notify the Revenue and Customs that he is in business and failure to do so may result in the imposition of £100 penalty. Either send a letter or use the form provided at the back of the *'Starting your own business booklet'* to the National Insurance Contributions Office and they will inform your local tax office of the change in your employment status.

Point of contact: telephone directory for address.

**National Insurance Contributions Office**

Self Employment Services
CAAT
Longbenton
Newcastle NE 98 1ZZ

Telephone the Call Centre on 0845 9154655 and ask for the following publications:

| | |
|---|---|
| CWL2 | Class 2 and Class 4 Contributions for the Self Employed |
| CA02 | People with Small Earnings from Self-Employment |
| CA04 | Direct Debit - The Easy Way to Pay. Class 2 and Class 3 |
| CA07 | Unpaid and Late Paid Contributions and for Employers |
| CWG1 | Employer's Quick Guide to PAYE and NIC Contributions |
| CA30 | Employer's Manual to Statutory Sick Pay |

## VAT

The VAT office also offer a number of useful publications, including;

| | |
|---|---|
| 700 | The VAT Guide |
| 700/1 | Should I be Registered for VAT? |
| 731 | Cash Accounting |
| 732 | Annual Accounting |
| 742 | Land and Property |

Information about the Cash Accounting Scheme and the introduction of annual VAT returns are dealt with later.
Point of contact: telephone directory for address.

### Local authorities

Authorities vary in provisions made for small businesses but all have been asked to simplify and cut delays in planning applications. In Assisted Areas, rent-free periods and reductions in rates may be available on certain industrial and commercial properties. As a preliminary to either purchasing or renting business premises, the following booklets will be helpful:

*Step by Step Guide to Planning Permission for Small Businesses,* and
  *Business Leases and Security of Tenure*

Both are issued by the Department of Employment and are available at council offices, Citizens' Advice Bureaux and TEC offices. Some authorities run training schemes in conjunction with local industry and educational establishments.

Point of contact: usually the Planning Department - ask for the Industrial Development or Economic Development Officer.

### Department of Business, Enterprise and Regulatory Reform (BERR)

The services formally provided by the Department are now increasingly being provided by Business Link . The Department can still, however, provide useful information on available grants for start-ups.
Point of contact: telephone 0207-215 5000 and ask for the address and telephone number of the nearest BERR office and copies of their explanatory booklets.

### Department of Transport and the Regions

Regulations are now in force relating to all forms of waste other than normal household rubbish. Any business that produces, stores, treats, processes, transports, recycles or disposes of such waste has a 'duty of care' to ensure it is properly discarded and dealt with.

Practical guidance on how to comply with the law (it is a criminal offence punishable by a fine not to) is contained in a booklet *Waste Management: The Duty of Care: A Code of Practice,* obtainable from HMSO Publication Centre, PO Box 276, London SW8 5DT. Telephone 0207-873 9090.

### Accountant

The services of an accountant are to be strongly recommended from the beginning because the legal and taxation requirements start immediately and must be properly complied with if trouble is to be avoided later. A qualified accountant must be used if a limited company is being formed but an accountant will give advice on a whole range of business issues including book-keeping, tax planning and compliance to finance raising and will help in preparing annual accounts.

It is worth spending some time finding an accountant who has other clients in the same line of business and is able to give sound advice particularly on taxation and business finance and is not so overworked that damaging delays in producing accounts are likely to arise. Ask other traders whether they can recommend their own accountant. Visit more than one firm of accountants, ask about the fees they charge and how much the production of annual accounts and agreement with the Revenue are likely to cost. A good accountant is worth every penny of his fees and will save you both money and worry.

**Solicitor**

Many businesses operate without the services of a solicitor but there are a number of occasions when legal advice should be sought. In particular, no-one should sign a lease of premises without taking legal advice because a business can encounter financial difficulty through unnoticed liabilities in its lease. Either an accountant or solicitor will help with drawing up a partnership agreement which all partnerships should have. A solicitor will also help to explain complex contractual terms and prepare draft contracts if the type of business being entered into requires them.

**Insurance broker**

Policies are available to cover many aspects of business including:

- employer's liability - compulsory if the business has employees
- public liability - essential in the construction industry
- motor vehicles
- theft of stock, plant and money
- fire and storm damage
- personal accident and loss of profits
- keyman cover.

Brokers are independent advisers who will obtain competitive quotations on your behalf. See more than one broker before making a decision - their advice is normally given free and without obligation.
Point of contact: telephone directory or write for a list of local members to:

The British Insurance Brokers' Association
  Consumer Relations Department
  BIBA House
  14 Bevis Marks
  London
  EC3A 7NT (telephone: 0207-623 9043)
                  or contact
  The Association of British Insurers
  51 Gresham Street
  London
  EC2V 7HQ (telephone: 0207-600 3333)

The British Insurance Brokers' Association
Consumer Relations Department
BIBA House
14 Bevis Marks
London
EC3A 7NT (telephone: 0207-623 9043)
or contact
The Association of British Insurers
51 Gresham Street
London
EC2V 7HQ (telephone: 0207-600 3333)

who will supply free a package of very useful advice files specially designed for the small business.

## The Health and Safety Executive

The Executive operates the legislation covering everyone engaged in work activities and has issued a very useful set of '*Construction Health Hazard Information Sheets*' covering such topics as handling cement, lead and solvents, safety in the use of ladders, scaffolding, hoists, cranes, flammable liquids, asbestos, roofs and compressed gases etc. A pack of these may be obtained free from your local HSE office or The Health & Safety Executive Central Office, Sheffield (telephone: (01142-892345) or HSE Publications (telephone: 01787-881165).

## Business plan

As stated before, once the relevant information has been obtained it should be consolidated into a formal business plan. The complexity of the plan will depend in the main on the size and nature of the business concerned. Consideration should be given to the following points.

Objectives

It is important to establish what you are trying to achieve both for you and the business. A provider of finance may be particularly influenced by your ability to achieve short- and medium-term goals and may have confidence in continuing to provide finance for the business. From an individual point of view, it is important to establish goals because there is little point in having a business that only serves to achieve the expectations of others whilst not rewarding the would-be businessman.

History

If you already own an existing business then commentary on its existing background structure and history to date can be of assistance. There is no substitute for experience and any existing contacts you have in the construction industry will be of assistance to you. The following points should also be considered for inclusion:

- a brief history of the business identifying useful contacts made
- the development of the business, highlighting significant successes and
  their relevance to the future
- principal reasons for taking the decision to pursue this new venture
- details of present financing of the business.

Products or services

It is important to establish precisely what it is you are going to sell. Does the product or service have any unique qualities which gives it your advantages over competitors? For example, do you have an ability to react more quickly than your competitors and are you perceived to deliver a higher quality product or service?
A typical business plan would include:

- description of the main products and services
- statement of disadvantages and advising how they will be overcome
- details of new products and estimated dates of introduction
- profitability of each product
- details of research and development projects
- after-sales support.

Markets and marketing strategy

This section of the business plan should show that thought has been given to the potential of the product. In this regard it can often be useful to identify major competitors and make an overall assessment of their strengths and weaknesses, including the following:

- an overall assessment of the market, setting out its size and growth potential
- a statement showing your position within the market
-

- an identification of main customers and how they compare
- details of typical orders and buying habits
- pricing strategy
- anticipated effect on demand of pricing
- expectation of price movement
- details of promotions and advertising campaigns.

It is important to identify your customers and why they might buy from you. Those entering the domestic side of the business will need to think about the best way to reach potential customers. Are local word-of-mouth recommendations enough to provide reasonable work continuity. If not, what is the most effective method of advertising to reach your customer base?

Remember, advertising is costly. It is a waste of funds to place an advertisement in a paper circulating in areas A, B, C & D if the business only covers area A.

Research and development

If you are developing a product or a particular service, then an assessment should be made on what stage it is at and what further finance is required to complete it. It may also be useful to make an assessment on the vulnerability of the product or service to innovations being initiated by others.

Basis of operation

Detail what facilities you will require in order to carry on your trade in the form of property, working and storage areas, office space, etc. An assessment should also he made on the assistance you will require from others.

Your business plan might include:

- a layman's guide to the process or work
- details of facilities, buildings and plant
- key factors affecting production, such as yields and wastage
- raw material demand and usage.

Management

This section is one of the most important because it demonstrates the capability of the would-be businessman. The skills you need will cover production, marketing, finance and administration. In the early stages you may be able to do this yourself but as the business grows it may be required to develop a team to handle these matters. The following points should be considered for inclusion in the plan:

- set out age, experience and achievements
- state additional management requirements in the future and how they are to be met
- identify current weaknesses and how they will be overcome
- state remuneration packages and profit expectations
- give detailed CVs in appendices.

Advertising and retraining may be required in order to identify and provide suitable personnel where expertise and experience are lacking.

Financial information

It is important to detail, if any, the present financial position of your business and the budgeted profit and loss accounts, cash flows and balance sheets. These integrated forecasts should be prepared for the next twelve months at monthly intervals and annually for the following two years.

If the forecasts are to be reasonably accurate then the businessman must make some early decisions about:

- the premises where the business will be based, the initial repairs and alterations that might he required and an assessment of the total cost
- which plant, equipment and transport are needed, whether they are to be leased or purchased and what the cost will be?
- how much stock of materials, if any, should be carried? - the bare minimum only should be acquired, so reliable suppliers should be found
- what will be the weekly bills for overheads, wages and the proprietor's living costs?
- what type of work is going to be undertaken, and how much profit can realistically be obtained?
- how often are invoices to be presented?

Your business plan should include the following information:

- explanation of how sales forecasts are prepared
- levels of production
- details of major variable overheads and estimates
- assumptions in cash flow forecasting, inflation and taxation.

Finance required and its application

The financial details given above should produce an accurate assessment of the funds required to finance the business. It is important to distinguish between those items that require permanent finance and those that will eventually be converted to cash because it is not usually advisable to finance long-term assets with personal equity.

Working capital such as stock and debtors can usually be obtained by an overdraft arrangement but your accountant or bank will advise you on this.

Executive summary

Although it is prepared last, this summary will be the first part of your business plan. Remember that business plans are prepared for busy people and their decision on finance may be based solely on this section. It should cover two or three pages and deal with the most important aspects and opportunities in your plan. Here are some of the main headings:

- key strategies
- finance required and how it is to be used
- management experience
- anticipated returns and profits
- markets.

The appendices should include:

- CVs of key personnel
- organisation charts
- market studies
- product advertising literature
- professional references
- financial forecasts
- glossary of terms.
- anticipated returns and profits
- markets.

If you feel that any additional information should be provided in support of your proposal, then this is usually best included in the appendices.

Follow up

Please remember that once your plan is prepared, it is important to re-examine it regularly and update the forecasts and financial information. This is a working document and can be an important tool in running the business.

## Sources of finance

Personal funds

Finance, like charity, often begins at home and a would-be businessman should make a realistic assessment of his net worth, including the value of his house after deducting the mortgage(s) outstanding on it, savings, any car or van owned and any sums which the family are prepared to contribute but deducting any private borrowings which will come due for payment. The whole of these funds may not be available (for instance, money which has been loaned to a friend or relative who is known to be unable to repay at the present time).

It may not be desirable that all capital should be put at risk on a business venture so the following should be established:

- how much cash you propose to invest in the business
- whether the family home will be made available for any business borrowing
- state total finance required
- how finance is anticipated being raised
- interest and security to be provided
- expected return on investment.

Whilst it may be wise not to pledge too much of the family assets, it has to be remembered that the bank will be looking closely at the degree to which the proprietor has committed himself to the venture and will not be impressed by an application for a loan where the applicant is prepared to risk only a small fraction of his own resources.

Having decided how much of his own funds to contribute, the businessman can now see the level of shortfall and consider how best to fill it. Consideration should be given to partners where the shortfall is large and particularly when there is a need for heavy investment in fixed assets, such as premises and capital equipment. It may be worthwhile starting a limited company with others also subscribing capital and to allow the banks to take security against the book debts.

Banks

The first outside source of money to which most businessmen turn is the bank and here are a few guidelines on approaching a bank manager:

- present your business plan to him; remember to use conservative estimates which tend to understate rather than overstate the forecast sales and profits
- know the figures in detail and do not leave it to your accountant to explain them for you. The bank manager is interested in the businessman not his advisers and will be impressed if the businessman demonstrates a grasp of the financing of his business
- understand the difference between short- and long-term borrowing
- ask about the Government Loan Guarantee Scheme if there is a shortage of security for loans. The bank may be able to assist, or depending on certain conditions being met, the Government may guarantee a certain percentage of the bank loan.

Remember the bank will want their money back, so bank borrowings are usually required to be secured by charges on business assets. In start-up situations, personal guarantees from the proprietors are normally required. Ensure that if these are given they are regularly reviewed to see if they are still required.

Enterprise Investment Scheme - business angels

If an outside investor is sought in a business he will probably wish to invest within the terms of the Enterprise Investment Scheme which enables him to gain income tax relief at 20% on the amount of his investment. Additionally, any investment can be used to defer capital gains tax. The rules are complex and professional advice should always be sought.

Hire purchase/leasing

It is not always necessary to purchase assets outright that are required for the business and leasing and hire purchase can often form an integral part of a business's medium-term finance strategy.

Venture capital

In addition, there are a number of other financial institutions in the venture capital market that can help well-established businesses, usually limited companies, who wish to expand. They may also assist well-conceived start-ups. They will provide a flexible package of equity and loan capital but only for large amounts, usually sums in excess of £150,000 and often £250,000.

Usually the deal involves the financial institution having a minority interest in the voting share capital and a seat on the board of the company. Arrangements for the eventual purchase of the shares held by the finance company by the private shareholders are also normally incorporated in the scheme.

The Royal Jubilee and Princes Trust

These trusts through the Youth Business Initiative provide bursaries of not more than £1,000 per individual to selected applicants who are unemployed and age 25 or over. Grants may be used for tools and equipment, transport, fees, insurance, instruction and training but not for working capital, rent and rates, new materials or stock. They operate through a local representative whose name and address may be ascertained by contacting the Prince's Youth Business.
Point of contact: telephone 0207-321 6500.

The Business Start-up Scheme

This is an allowance of £50 per week, in addition to any income made from your business, paid for twenty weeks. To qualify you must be at least 18 and under 65, work at least 36 hours per week in the business and have been unemployed for at least six months or fall into one of the other categories: disabled, ex-HMS or redundant.

The first step is to get the booklet on the subject from your local Jobcentre or TEC that includes details on how and where to apply. Once in receipt of the enterprise allowance, you will also have the benefit of advice and assistance from an experienced businessman from your TEC. All the initial counselling services and training courses are free.

# RUNNING A BUSINESS

Many businesses are run without adequate information being available to check trend in their vital areas, e.g. marketing, money and managerial efficiency. It is essential to look critically at all aspects of the business in order to maximise profits and reduce inefficiency. Regular meaningful information is required on which management can concentrate. This will vary according to the proprietor's business but will often concentrate on debtors, creditors, cash, sales and orders.

Proprietors often have the feeling that the business should be 'doing better' but are unable to identify what is going wrong. Sometimes there is the worrying phenomenon of a steadily increasing work programme coupled with a persistently reducing bank balance or rising overdraft. Some useful ways of checking the position and of identifying problem areas are given below.

## Marketing

Throughout his business life the entrepreneur should continuously study the methods and approach of his competitors. A shortcoming frequently found in ailing concerns is that the proprietor thinks he knows what his customers want better than they do.

The term 'market research' sounds both difficult and expensive but a very simple form of it can be done quite effectively by the businessman and his sales staff. Existing and prospective customers should be approached and asked what they want in terms of price, quality, design, payment terms, follow-up service, guarantees and services.

The initial approach might be by a leaflet or letter followed by a personal call. As an on-going part of management, all staff with customer contact should be encouraged to enquire about and record customer preferences, complaints, etc. and feed it back to management.

Other sources of information can be trade and business journals, trade exhibitions, suppliers and representatives from which information about trends, new techniques and products can be obtained and studied. Valuable information can also be gained from studying competitors and the following questions should be asked:

- what do they sell and at what prices?
- what inducements do they offer to their customers, e.g. credit facilities, guarantees, free offers and discounts?
  how do they reach their customers - local/national advertising, mail shots, salesmen, local radio and TV?
- what are the strongest aspects of their appeal to customers and
- have they any weaknesses?

The businessman should apply all the information gathered from customers and competitors to his own services with a view to making sure he is offering the right product at the right price in the most attractive way and in the most receptive market.

In a small business where the proprietor is also his own salesman he must give careful thought on how he can best present his product and himself. For instance, if he is working solely within the construction industry his main problems are likely to centre on getting a C1S6 Certificate and using trade contacts to get sub-contract work.

However, for those who serve the general public, presentation can be a vital element in getting work. The customer is looking for efficiency, reliability and honesty in a trader and quality, price and style in the product. To bring out these facets in discussion with a potential customer is a skilled task. A short course on marketing techniques could pay handsome dividends. The Business Link will give the names and addresses of such courses locally.

## Financial control

Unfortunately, some unsuccessful firms do not seek financial advice until too late when the downward trend cannot be halted. Earlier attention to the problems may have saved some of them so it is important to recognise the tell-tale signs. There are some tests and checks that can be done quite easily.

## Cash flow

Cash flow is the lifeblood of the business and more businesses fail through lack of cash than for any other reason. Cash is generated through the conversion of work into debtors and then into payment and also through the deferral of the payment of supplies for as long a period that can be negotiated. The objective must be to keep stock, work in progress, debts to a minimum and creditors to a maximum.

## Debtor days

This is calculated by dividing your trade debtors by annual sales and multiplying by 365. This shows the number of days' credit being afforded to your customers and should be compared both with your normal trade terms and the previous month's figures. Normal procedures should involve the preparation of a monthly-aged list of debtors showing the name of the customer, the value and to which month it relates.

The oldest and largest debtors can be seen at a glance for immediate consideration of what further recovery action is needed. The list may also show over-reliance on one or two large customers or the need to stop supplying a particularly bad payer until his arrears have been reduced to an acceptable level. Consideration should be given to making up bills to a date before the end of the month and making sure the accounts are sent out immediately, followed by a statement four weeks later.

Consider giving discounts for prompt payment. If all else fails, and legal action for recovery is being contemplated, call at the County Court and ask for their leaflets.

## Stock turn

The level of stock should be kept to a minimum and the number of days' stock can be calculated by dividing the stock by the annual purchases and multiplying by 365. A worsening trend on a month-by-month basis shows the need for action. It is important to regularly make a full inventory of all stock and dispose of old or surplus items for cash. A stock control procedure to avoid stock losses and to keep stock to a minimum should be implemented.

## Profitability

Whilst cash is vital in the short-term, profitability is vital in the medium-term. The two key percentage figures are the gross profit percentage and the net profit percentage. Gross profit is calculated by deducting the cost of materials and direct labour from the sales figures whilst net profit is arrived at after deducting all overheads. Possible reasons for changes in the gross profit percentage are:

- not taking full account of increases in materials and wages in the pricing of jobs
- too generous discount terms being offered
- poor management, over-manning, waste and pilferage of materials
- too much down-time on equipment which is in need of replacement.

If net profit is deteriorating after the deduction of an appropriate reward for your own efforts, including an amount for your own personal tax liability, you should review each item of overhead expenditure in detail asking the following questions:

- can savings be made in non-productive staff?
- is sub-contracting possible and would it be cheaper?

- have all possible energy-saving methods been fully explored?
- do the company's vehicles spend too much time in the yard and can they be shared or their number reduced?
- is the expenditure on advertising producing sales - review in association with 'marketing' above?

## Over-trading

Many inexperienced businessmen imagine that profitability equals money in the bank and in some cases, particularly where the receipts are wholly in cash, this may be the case. But often, increased business means higher stock inventories, extra wages and overheads, increased capital expenditure on premises and plant, all of which require short-term finance.

Additionally, if the debtors show a marked increase as the turnover rises, the proprietor may find to his surprise that each expansion of trade reduces rather than increases his cash resources and he is continually having to rely on extensions to his existing credit.

The business, which had enough funds for start-up, finds it does not have sufficient cash to run at the higher level of operation and the bank manager may he getting anxious about the increasing overdraft. It is essential for those who run a business that operates on credit terms to be aware that profitability does not necessarily mean increased cash availability. Regular monthly management information on marketing and finance as described in this chapter will enable over-trading to be recognised and remedial action to be taken early.

If the situation is appreciated only when the bank and other creditors are pressing for money, radical solutions may be necessary, such as bringing in new finance, sale and leaseback of premises, a fundamental change in the terms of trade or even selling out to a buyer with more resources. Professional help from the firm's accountant will be needed in these circumstances.

## Break-even point

The costs of a business may be divided into two types - variable and fixed. *Variable costs* are those which increase or decrease as the volume of work goes up or down and include such items as materials used, direct labour and power machine tools. *Fixed costs* are not related to turnover and are sometimes called fixed overheads. They include rent, rates, insurance, heat and light, office salaries and plant depreciation. These costs are still incurred even though few or no sales are being made.

Many small businessmen run their enterprises from home using family labour as back-up; they mainly sell their own labour and buy materials and hire plant only as required. By these means they reduce their fixed costs to a minimum and start making profits almost immediately. However, larger firms that have business premises, perhaps a small workshop, an office and vehicles, need to know how much they have to sell to cover their costs and become profitable.

In the case of a new business it is necessary to estimate this figure but where annual accounts are available a break-even chart based on them can be readily prepared. Suppose the real or estimated figures (expressed in £000s) are:

|                | %   | £   |
|----------------|-----|-----|
| Sales          | 100 | 400 |
| Variable costs | 66  | 265 |
| Gross profit   | 34  | 135 |
| Fixed costs    | 13  | 50  |
| Net profit     | 21  | 85  |

$$\text{Break-even point} = \frac{50}{1 \text{ less variable costs \%}} \text{ sales}$$

$$= 50 \text{ divided by } (1 \text{ less } 0.6625)$$
$$= 50 \text{ divided by } 0.3375$$
$$= £148 \text{ (thousand)}$$

In practice things are never quite as clear cut as the figures show, but nevertheless this is a very useful tool for assessing not only the break-even point but also the approximate amount of loss or profit arising at differing levels of turnover and also for considering pricing policy.

# TAXATION

The first decision usually required to be made from a taxation point of view is which trading entity to adopt. The options available are set out below.

## Sole trader

A sole trader is a person who is in business on his own account. There is no statutory requirement to produce accounts nor is there a necessity to have them audited. A sole trader may, however, be required to register for PAYE and VAT purposes and maintain records so that Income Tax and VAT returns can be made. A sole trader is personally liable for all the liabilities of his business.

## Partnership

A partnership is a collection of individuals in business on their own account and whose constitution is generally governed by the Partnership Act 1890. It is strongly recommended that a partnership agreement is also established to determine the commercial relationship between the individuals concerned.

The requirements in relation to accounting records and returns are similar to those of a sole trader and in general a partner's liability is unlimited.

## Limited company

This is the most common business entity. Companies are incorporated under the Companies Act 1985 which requires that an annual audit is carried out for all companies with a turnover in excess of £1,000,000 or a review if the turnover is less than £1,000,000 and that accounts are filed with the Companies Registrar. Generally an individual shareholder's liability is limited to the amount of the share capital he is required to subscribe.

## Advantages

In view of the problems and costs of incorporating an existing business, it is important to try and select the correct trading medium at the commencement of operations. It is not true to say that every business should start life as a company.

Many businesses are carried on in a safe and efficient manner by sole traders or partnerships. Whilst recognising the possible commercial advantages of a limited company, taxation advantages exist for sole traderships and partnerships, such as income tax deferral and National

Insurance saving. No decision should be taken without first seeking professional advice.

The benefit of limited liability should not be ignored although this can largely be negated by banks seeking personal guarantees. In addition, it may be easier for the companies to raise finance because the bank can take security on the debts of the company that could be sold in the future, particularly if third-party finance has been obtained in the form of equity.

## Self-assessment

From the tax year 1996/97 the burden of assessing tax shifted from the Revenue to the individual tax payer. The main features of this system are as follows:

- the onus is on the taxpayer to provide information and to complete returns
- tax will be payable on different dates
- the taxpayer has a choice: he can calculate his tax liability at the same time as making his return and this will need to be done by 31 January following the end of the tax year. Alternatively, he can send in his tax return before 30 September and the Revenue will calculate the tax to be paid on the following 31 January
- the important aspect to the system is that if the return is late, or the tax is paid late, there will be automatic penalties and/or surcharges imposed on the taxpayer.

## Tax correspondence

Businessmen do not like letters from the Revenue but they should resist the temptation to tear them up or put them behind the clock and forget about them. All Tax Calculations and Statements of Account should be checked for accuracy immediately and any queries should be put to your accountant or sent to the Tax District that issued the document.

Keep copies of all correspondence with the Revenue. Letters can be mislaid or fail to be delivered and it is essential to have both proof of what was sent as well as a permanent record of all correspondence.

## Dates tax due

*Income Tax*

Payments on account (based on one half of last year's liability) are due on 31 January and 31 July. If these are insufficient there is a balancing payment due on the following 31 January – the same day as the tax return needs to be filed. For example:

for the year 2007/08    Tax due £5,000 (2006/07 was £4,000)
First payment on account of £2,000 is due on 31.01.08
Second payment on account of £2,000 is due on 31.07.08
Balancing payment of £1,000 is due on 31.01.09

Note that on 31.01.08, the first payment on account of £2,500 falls due for the tax year 2008/09.

**Tax in business**

*Spouses in business*

If spouses work in the business, perhaps answering the phone, making appointments, writing business letters, making up bills and keeping the books, they should be properly remunerated for it. Being a payment to a family member, the Inspector of Taxes will be understandably cautious in allowing remuneration in full as a business expense. The payment should be:

- actually paid to them, preferably weekly or monthly and in addition to any housekeeping monies
- recorded in the business book
- reasonable in amount in line with their duties and the time spent on them.

If the wages paid to them exceed £104.00 per week, Class 1 employer's and employee's NIC becomes due and if they exceed £5,435 p.a. (assuming they have no other income) PAYE tax will also be payable.

It should also be noted that once small businesses are well established and the spouses' earnings are approaching the above limits, consideration may be given to bringing them in as a partner. This has a number of effects:

- there is a reduced need to relate the spouse's income (which is now a share of the profits) to the work they do
- they will pay Class 2 and Class 4 NIC instead of the more costly Class I contributions and PAYE will no longer apply to their earnings but remember that, as partners, they have unlimited liability.

## Premises

Many small businessmen cannot afford to rent or buy commercial premises and run their enterprises from home using part of it as an office where the books and vouchers, clients' records and trade manuals are kept and where estimates and plans are drawn up. In these circumstances, a portion of the outgoings on the property may be claimed as a business expenses. An accountant's advice should be sought to ensure that the capital gains tax exemption that applies on the sale of the main residence is not lost.

## Fixed Profit Car Scheme

It may be advantageous to calculate your car expenses using a fixed rate per business mile. A condition is that your annual turnover is below the VAT threshold (currently £60,000). Ask your accountant about this. A proper record of business mileage must be kept.

## Vehicles

Car expenses for sole traders and partners are usually split on a fractional mileage basis between business journeys, which are allowable, and private ones, which are not, and a record of each should he kept. If the business does work only on one or two sites for only one main contractor, the inspector may argue that the true base of operations is the work site not the residence and seek to disallow the cost of travel between home and work. It is tax-wise and sound business practice to have as many customers as possible and not work for just one client.

## Business entertainment

No tax relief is due for expenditure on business entertainment and neither is the VAT recoverable on gifts to customers, whether they are from this country or overseas. However, the cost of small trade gifts not exceeding £10 per person per annum in value is still admissible provided that the gift advertises the business and does not consist of food, drink or tobacco.

### Income tax (2008/09)

*Personal allowances*

The current personal allowance for a single person is £5,435. The personal allowance for people aged 65 to 74 and over 75 years are £9,030 and £9,180 respectively. The married couple's allowance was withdrawn on 5 April 2000, except for those over 65 on that date.

Taxation of husband and wife

A married woman is treated in much the same way as a single person with her own personal allowance and basic rate band. Husband and wife each make a separate return of their own income and the Inland Revenue deals with each one in complete privacy; letters about the husband's affairs will be addressed only to him and about the wife's only to her unless the parties indicate differently.

## Rates of tax

Tax is deducted at source from most banks and building societies accounts at the rate of 20%. The rates of tax on earnings for 2008/09 are as follows:

Basic rate: 20% on taxable income between £2,231 and £34,600
Higher rate: 40% on taxable income over £34,600

Dividends carry a 10% non-repayable tax credit. Higher rate taxpayers pay a further tax on dividends of 22.5%.

## Mortgage interest relief

This is no longer available after 5 April 2000.

## Business losses

These are allowed only against the income of the person who incurs the loss. For example, a loss in the husband's business cannot be set against the wife's income from employment.

## Joint income

In the case of joint ownership by a husband and wife of assets that yield income, such as bank and building society accounts, shares and rented property, the Inland Revenue will treat the income as arising equally to both and each will pay tax on one half of the income. If, however, the asset is owned in unequal shares or one spouse only and the taxpayer can prove this, then the shares of income to be taxed can be adjusted accordingly if a joint declaration is made to the tax office setting out the facts.

**Capital Gains Tax**

Where an asset is disposed of, the first £9,200 of the gain is exempt from tax. In the case of husbands and wives, each has a £9,200 exemption so if the ownership of the assets is divided between them, it is possible to claim exemption on gains up to £18,400 jointly in the tax year. Any remaining gain is chargeable at 18%.

**Self-employed NIC rates (from 6 April 2008)**

*Class 2 rate*
Charged at £2.30 per week. If earnings are below £4,825 per annum averaged over the year, ask the Revenue about 'small income exception'. Details are in leaflet CA02.

*Class 4 rate*
Business profits up to £5,435 per annum are charged at NIL. Annual profits between £5,435 and £40,040 are charged at 8% of the profit. There is a charge on profits over £40,040. Class 4 contributions are collected by the IRevenue along with the income tax due.

*Capital allowances (depreciation) rates*

| | |
|---|---|
| Plant and machinery: | 25% (50% first-year allowance is available for certain small businesses) |
| Business motor cars - cost up to £12,000: | 25% |
|                - cost over £12,000: | £3,000 (maximum) |

## THE CONSTRUCTION INDUSTRY TAX DEDUCTION SCHEME

**General**

The new Construction Industry Tax Deduction Scheme is known as the 'CIS' scheme and replaced the old '714' scheme. As the scheme operates whenever a contractor makes a payment to a sub-contractor, the businessman should visit his local income tax enquiry office and obtain copies of the Inland Revenue booklet IR 14/15 (CIS) and leaflet IR 40 which will explain the conditions under which the Inland Revenue will issue a registration card or (CIS6) certificate and precisely when the scheme applies.

Everyone who carries out work in the Construction Industry Scheme must hold a registration card (CIS4) or a tax certificate (CIS6). Certain larger companies use a special certificate (CIS5).

If the sub-contractor has a registration card but does not hold a valid tax certificate (CIS6) issued to him by the Inland Revenue, then the contractor *must* deduct 18% tax from the whole of any payment made to him (excluding the cost of any materials) and to account to the Inland Revenue for all amounts so withheld.

To enable the subcontractor to prove to the Inspector of Taxes that he has suffered this tax deduction, the contractor must complete the three-part tax payment voucher (CIS25) showing the amount withheld. These vouchers must be carefully filed for production to the Inspector after the end of the tax year along with the tax return. Any tax deducted in this way over and above the sub-contractor's agreed liability for the year will be repaid by the Inland Revenue. If he holds a (CIS6) certificate the payment may be made in full without deducting tax.

A small business that does work only for the general public and small commercial concerns is outside the scheme and does not need a certificate to trade. If, however, it engages other contractors to do jobs for it, the business would have to register under the scheme as a contractor and deduct tax from any payment made to a sub-contractor who did not produce a valid (CIS6) certificate. If in doubt, consult your accountant or the Inland Revenue direct.

## VAT

The general rule about liability to register for VAT is given in the VAT office notes. It is possible to give here only a brief outline of how the tax works. The rules that apply to the construction industry are extremely complex and all traders must study *The VAT Guide* and other publications.

Registration for VAT is required if, at the end of any month, the value of taxable supplies in the last 12 months exceeds the annual threshold or if there are reasonable grounds for believing the value of the taxable supplies in the next 30 days will exceed the annual threshold.

Taxable supplies include any zero-rated items. The annual threshold is £60,000. The amount of tax to be paid is the difference between the VAT charged out to customers *(output tax)* and that suffered on payments made to suppliers for goods and services *(input tax)* incurred in making taxable supplies. Unlike income tax there is no distinction in VAT for capital items so that the tax charged on the purchase of, for example, machinery, trucks and office furniture, will normally be reclaimable as *input tax*.

VAT is payable in respect of three monthly periods known as 'tax periods'. You can apply to have the group of tax periods that fits in best with your financial year. The tax must be paid within one month of the end of each tax period. Traders who receive regular repayments of VAT can apply to havthem monthly rather than quarterly. Not all types of goods and services are taxed at 17.5% (i.e. the standard rate). Some are exempt and others are zero-rated.

## Zero-rated

This means that no VAT is chargeable on the goods or services, but a registered trader can reclaim any *input* tax suffered on his purchases. For instance, a builder pays VAT on the materials he buys to provide supplies of constructing but if he is constructing a new dwelling house, this is zero rated. The builder may reclaim this VAT or set it off against any VAT due on standard rated work.

## Exempt

Supplies that are exempt are less favourably treated than those that are zero rated. Again no VAT is chargeable on the goods or services but the trader cannot reclaim any *input* tax suffered on his purchases.

## Standard-rated

All work which is not specifically stated to be zero rated or exempt is standard-rated, i.e. VAT is chargeable at the current rate of 17.5% and the trader may deduct any *input* tax suffered when he is making his return to the Customs and Excise. If for any reason a trader makes a supply and fails to charge VAT when he should have done so (e.g. mistakenly assuming the supply to be zero rated), he will have to account for the VAT himself out of the proceeds. If there is any doubt about the VAT position, it is safer to assume the supply is standard rated, charge the appropriate amount of VAT on the invoice and argue about it later.

## Time of supply

The *time* at which a supply of goods or services is treated as taking place is important and is called the 'tax point'. VAT must be accounted for to the Customs and Excise at the end of the accounting period in which this 'tax point' occurs. For the supply of goods which are 'built on site', the 'basic tax point' is the date the goods are made available for the customer's use, whilst for *services* it is normally the date when all work except invoicing is completed.

However, if you issue a tax invoice or receive a payment before this 'basic tax point' then that date becomes a tax point. In the case of contracts providing for stage and retention payments, the tax point is either the date the tax invoice is issued or when payment is received, whichever is the earlier.

All the requirements apply to sub-contractors and main contractors and it should be noted that, when a contractor deducts income tax from a payment to a sub-contractor (because he has no valid CIS6) VAT is payable on the full gross amount *before* taking off the income tax.

## Annual accounting

It is possible to account for VAT other than on a specified three month period. Annual accounting provides for nine equal installments to be paid by direct debit with annual return provided with the tenth payment. £300,000.

## Cash accounting

If turnover is below a specified limit, currently £600,000, a taxpayer may account for VAT on the basis of cash paid and received. The main advantages are automatic bad debt relief and a deferral of VAT payment where extended credit is given.

## Bad debts

Relief is available for debts over 6 months.

# Index

Access and site equipment,
  244-246
Accountant, 275
Alterations and repairs,
  173-196
Aluminium
  gutters, 104
  pipes, 100
Approximate estimating,
  227-240
Asphalt generally
  damp-proofing, 75-76
  flooring, 76-77
  roofing, 77-78
  tanking, 75-76
 Asphalt work, 75-78

Balustrades, 60
Banks, 273
Basins, 92-93
Bath accessories, 91-92
Baths, 90-91
Bearers, 45-46
Bidets, 94
Bitumen macadam paving, 171
Blockwork, 30-34
Boilers
  gas, 120-121
  oil, 122
Breaking and demolition,
  246
Break-even point, 290-291
Brick paving, 172
Brickwork, 20-30
Brickwork and blockwork,
  20-36, 182-185, 229-230,
  261-262
Built up roofing, 73-74
Business links, 272
Business matters, 271-301
Business plan, 277-282

Capital Gains Tax, 298

Carpentry and Joinery, 41-59,
  189-191, 231-234, 262
Cashflow, 288
Cast iron
  gutters, 105
  rainwater pipes, 102-103
Cavity wall insulation, 36
Ceramic tiling, 88-90
Chainlink fencing, 166-167
Chestnut fencing, 167
Clear float glass, 132-133
Clear laminated glass, 132-133
Close boarded fencing, 167
Cold water storage tanks, 118-119
Concrete and cutting equipment,
  243-244
Concrete beds, 170-171
Concrete finishes, 18
Concrete mixes, 260
Concrete work, 11-19, 228
Construction Industry
  Deduction Scheme, 298-299
Contracting, xv
Copings, 19
Copper pipework
  capillary joints, 113-117
  compression joints, 117-118
Copper sheet coverings, 72-73
Cylinders, 119

Damp proof courses, 34-35
Damp proofing, 75-76
Debtor days, 288-289
Defects liability period, xvi
Demolition, 3-4
Demolition, excavation and filling,
  3-10
Department of Business,
Enterprise and Regulatory
  (BERR), 275
Department of Transport
  and the Regions, 275
Disposal, 7-8

Doors, 52-53
Door frames and linings, 53-55
Drainage, 157-163, 238-240

Earthwork support, 7
Emergency measures, 199-200
Estimates, viii
Excavation and filling, 5-6, 227,
   258-259
External works,157-172
   266-267

Fascias, 45
Fencing
   chainlink, 166-167
   chestnut, 167
   close boarded, 167
   strained wire, 167
Fibre-cement slating, 65-66
Filling, 9-10
Filling openings, 176
Finance
   banks, 283
   Business Start-up Scheme, 284-
      285
   Enterprise Investment
   Scheme, 283
   hire purchase/leasing, 284
   personal funds, 282-283
   Royal Jubilee and Princes
      Trust, 284
   venture capital, 284
Finishings, 79-89, 191-192,
   224
Fire damage, 201-211
Fixed Profit Car Scheme, 296
Flood damage, 212-213
Flooring, 48-49, 76-77
Floor tiling, 86-88
Floor, wall and ceiling finishings,
   79-89
Forming openings, 174-175
Formwork, 13-16
Fuel consumption, 259

Gale damage, 214-217
Gas boilers, 120-121
General construction data, 235-
   267
Georgian glass
   wired, 130-131
   wired polished, 131-132
Glass
   clear float, 126-128
   clear laminated safety, 132-133
   Georgian wired, 130-131
   Georgian wired polished,
      131-132
   white patterned, 128-130
Glazing, 126-133, 194
Granolithic screed, 81-82
Gravel paving, 171
Gutter boards, 44-45

Health and Safety
   Executive, 277
Hardwood windows, 51
Holes and chases, 97-99
Hot water cylinders, 199

Insulation, 57, 120
Insurance broker, 276, 277
Insurances, xvii
Ironmongery, 57-59

Joinery, (see Carpentry and
   Joinery)
Joints, 18

Kerbs and bearers, 42, 43-44
Kerbs and edgings, 168-170
Kitchen fittings, 56

Labour, xiv
Lathing, 85
Lead sheet coverings, 69-71
Lifting and moving, 246
Limited company, 293
Linings, 49-50

Lintels
  concrete, 18
  steel, 60

Manholes, 164-166, 240
Marketing, 287-288
Masonry, 37-40, 230-231, 262
Materials, xiv
Metal fixings, 46-47
Metalwork, 60-61

National Insurance
  Contributions, 273-274, 298
Natural slating, 66-68

Oil boilers, 122
Oil storage tanks, 122-123
Overflows, 9112
Overheads and profit, xiv
Over-trading, 290

Painting,
  internally, 134-149
  externally, 145-154
  generally, 134-154, 195, 236-
    237, 265
Partnership, 293
Pavings
  bitumen macadam, 171
  brick, 172
  gravel, 171
  precast concrete, 172
Payments, xv-xvi, xix
Plant and Tool Hire, 243-248
Plasterboard, 86
Plasterwork, 83-84, 264
Plumbing and heating, 90-125,
  192,194, 234-236, 264
Power tools, 247-248
Pointing, 36
Precast concrete
  copings, 19
  kerbs, 168-170
  lintels, 18
  pavings, 172

Premises, 296
Profitability, 289-290
Pumping equipment, 2348
PVC-U
  rainwater gutters,103
  rainwater pipes, 100
  windows, 51

Quarry tiling, 86-88
Quotations, xvi

Radiators, 123-125
Rainwater goods, 100-106
Rainwater gutters
  aluminium, 104
  cast iron, 105
  PVC-U, 103
Rainwater pipes
  aluminium, 100
  cast iron, 102-103
  PVC-U, 100
Reconstructed slating, 68-69
Reinforcement
  bars, 17
  fabric, 17
Retention, xv-xvi
Revenue and Customs, 273
Roofing, 186-189, 262-264,
Roof outlets, 106
Roof tiling, 62-65
Rubble walling, 37-38
Running a business, 287-291

Sanitary fittings, 90-96
Screeds
  cement and sand, 79-81
  granolithic, 81-82
Screens, 185
Sheet coverings
  copper, 72-73
  lead, 69-71
Shoring, 173
Showers, 96
Softwood windows, 50-51
Sole trader, 293

Slating
fibre-cement, 65-66
natural, 66-68
reconstructed, 68-69
Solicitor, 276
Stairs, 55-56
Steel lintels, 60
Stop valves, 104, 106
Strained wire fencing, 167
Strutting, 42, 44
Surface treatments, 10

Tanking, 75-76
Taps, 95-96
Taxation
allowances, 296-297
business, 295
correspondence, 294
dates due, 294-295
rates, 297
Temporary screens, 182
Theft damage, 218-213
Tiling
floor, 86-88
roof, 62-65
wall, 88-90

Tools and equipment hire,
243-248
Training and Enterprise Councils
272

Underpinning, 177-182
Urinals, 95

Valves, 96
Variations, xvi, xix-xx
VAT, xiii, 274, 299-301
Vehicles, 296
Vinyl tiling, 87-88

Wallpapering, 155-156,
195-196, 237-238, 265-266,
Wall tiling, 88-90
Waste pipes, 107-112
Water bars, 61
WCs, 94
Welding and generators, 248
Weights of materials, 254-258
White patterned glass, 128-130
Windows
hardwood, 51
PVC-U, 51
softwood, 50-51